Volume 1

Teaching
Ma²+(h)
with
Google Apps
50 G Suite Activities

Alice Keeler and Diana Herrington

foreword by Jo Boaler

Teaching Math with Google Apps

© 2017 by Alice Keeler and Diana Herrington

This book is available at special discounts when purchased in quantity for use as premiums, promotions, fundraising, and educational use. For inquiries and details, contact the publisher at shelley@daveburgessconsulting.com.

Google and the Google logo are registered trademarks of Google Inc.

Published by Dave Burgess Consulting, Inc.
San Diego, CA
http://daveburgessconsulting.com

Cover Design by Genesis Kohler
Interior Design by My Writers' Connection

Library of Congress Control Number: 2017935673
Paperback ISBN: 978-1-946444-04-2
eBook ISBN: 978-1-946444-05-9
First Printing: March 2017

Contents

Foreword by Jo Boaler

What a treat for any mathematics teacher and their students—a book filled with recipes for student engagement and learning. Inside the pages of this book you will find a lovely combination of rich tasks that invite students into important mathematical discoveries, and pedagogical advice that, if followed, will move students from passive receivers of knowledge to active inquirers.

We are in the twenty-first century, but visitors to many math classrooms could be forgiven for thinking they had stepped back in time and walked into the Victorian era. Students often spend a lot of their mathematics classroom time watching a teacher work on mathematics, then trying to follow their steps. Chalkboards may have been replaced with "smart boards" but very little else has changed.

Research tells us there is a different, more productive role for students in classrooms— one in which they are actively engaged learning mathematics, thinking about and discussing ideas, and collecting information and resources from many different places, not just the hand of the teacher. The authors of this book—two mathematics visionaries,

Alice Keeler and Diana Herrington—make an important point: If students can look up a solution to a question on the Internet, the question is not worth asking. Instead of banning technology for their students— which makes learning mathematics into an unworldly event that is unlike the rest of students' lives—these teacher-authors embrace it and give students more productive mathematical activities to think about and work on.

This book is filled with pedagogical ideas and advice, which are valuable, and it also provides a range of incredible tasks that invite students to be mathematical problem solvers and inquirers. We know from new brain evidence that our brains want to think visually about mathematics. When we work on mathematics, even an abstract calculation, five brain areas are involved and two of them are visual pathways (youcubed.org/visual-math/). For many students, their mathematics education is almost entirely numerical, giving them little or no opportunity to strengthen important pathways.

Increased mathematical understanding comes with increased communication between different areas of the brain. The activities in this book provide important brain communication opportunities as students work with both visuals and numbers. The activities also invite students to be problem solvers and inquirers, which are the most important roles for students. We don't know what mathematics students will need in their future lives; technology is changing at such a fast pace that most of what the mathematics students are learning in school now will be obsolete when they enter the workplace. In the future, we know we will need people who can problem solve, think quantitatively about the world, and reason with their ideas. Teachers have a choice: to teach students content passively so they only learn to reproduce steps, or to teach them through active engagement so that when they learn mathematics content they also learn to use it, to think creatively, and to apply ideas to different situations. The former is uninteresting for most students; the latter is engaging and important. This book provides the activities that invite students to think quantitatively, to use important technology, and to give students a learning experience that will help them develop an active identity as a mathematics problem solver.

Thirty years ago, the Fortune 500 companies in the United States were asked what they most valued from school graduates. The list on the following page shows calculation was second highest in demand. Ten years ago the list had completely changed, calculation was now second from bottom, and the most valued attributes became teamwork and problem solving. Employers will never need students to calculate, as we have technological tools that can do this more effectively than any human, but they will need students to reason, to think quantitatively, and to model with mathematics. Through the pages of this book, students will learn important mathematics, they will also learn to use it effectively.

Mathematics classrooms are often speed-driven spaces in which many students decide they are not a "math person" and give up. The source of this unproductive idea is often the speedy thinkers that drive the pace of classroom instruction. We know that many mathematics students, including those who have the potential to be world-changing mathematicians, think slowly and deeply about mathematics. When teachers ask questions and then take an answer from the first student that shoots up their hand, they give a very clear message: speed is what is valued in their classroom. Other

Fortune 500 most valued skills

30 years ago	10 years ago
Writing	Teamwork
Computational Skills	Problem Solving
Reading Skills	Interpersonal Skills
Oral Communications	Oral Communications
Listening Skills	Listening Skills
Personal Career Development	Personal Career Development
Creative Thinking	Creative Thinking
Leadership	Leadership
Goal Setting/Motivation	Goal Setting/Motivation
Teamwork	Writing
Organizational Effectiveness	Organizational Effectiveness
Problem Solving	Computational Skills
Interpersonal Skills	Reading Skills

From *Mathematical Mindsets*, by Jo Boaler, 2016

students who think less quickly, but just as well, often give up, thinking that they cannot be successful in math. This is a problem for any teacher. So how do teachers set the pace of a classroom when students think at different rates? Keeler and Herrington provide a perfect solution: they ask students to submit answers in a Google form. Speed is no longer an issue and student thinking is what is valued. Teachers can see the thinking of all the students, not just a select few, and students have interacted with an important piece of technology which they can appreciate and learn from.

This is just one of the many ideas in this important book that I am excited to be a part of. Invite students into the twenty-first century in your mathematics classroom, engage them with inspiring visual tasks, and equip them with the mathematical tools that they can use to solve problems and to make a difference in the world.

And, *viva la* revolution!

Jo Boaler

Stanford Professor,
co-founder of youcubed.org,
author of *Mathematical Mindsets*

Get to Know Alice

I have taught math 1:1 almost my entire teaching career. I started teaching math in 1999 after graduating with my BA in mathematics. While I often hear "you cannot teach math with technology," I truly do not know how to teach math without it. In 2011, I became a Google Certified Teacher (now Google Certified Innovator). I love helping educators almost as much as I enjoy working with students, which is why my blog, alicekeeler.com, provides EdTech help for teachers who want to learn more about using Google Apps. It's also the reason I have worked with Google on projects such as YouTube for Teachers and Google Play for Education and co-authored the books *50 Things You Can Do with Google Classroom* and *50 Things to Go Further with Google Classroom*.

Although I earned a bachelor's degree in math, I haven't always felt good about the subject. I am not bad at math, but like a lot of students, I struggled with rote memorization. In fact, I struggled so much with timed math tests in elementary school that I ended up repeating the fourth grade. I never did pass the timed tests, but I eventually learned to love math. Today, I can make killer spreadsheets to solve almost any problem (docs.googleblog.com/2015/06/meet-alice-keeler-google-certified.html), and I enjoy coding Google Apps Script.

I passionately believe that kids are not failures. When they don't do well in a subject, sometimes they may need a different approach to feel successful.

Get to Know Diana

I have been teaching math for four decades, and I still enjoy every minute of it. Of the student- and peer-nominated awards I've been honored to receive, the three I am most proud of are the Presidential Award for Excellence in Mathematics and Science Teaching (PAEMST), the California State Science Fair Coach of the Year (as the first non-science teacher recipient), and the Central Valley Computer Using Educator of the Year (CVCUE). I am also an alumna of the California Teacher Advisory Council (CalTAC) which is part of the California Council of Science and Technology (CCST), and have worked on several California statewide testing programs as a writer and a chief reader, including the Golden State Exam and the CLAS test.

I earned both a bachelor's and master's in mathematics from California Polytechnic State University, San Luis Obispo. The school, with its motto, "learn by doing," was a perfect fit for me. While at Cal Poly, mathematical modeling became one of my favorite activities. I also enjoyed the learning opportunities at a local military base with NASA. From those experiences, I learned that there are times when you create a model but you may not have the science to fully understand what you can do. Those college experiences also helped deepen my engagement in mathematics and gave me more things to be curious about. The more I learned and experienced, the more I saw mathematics in the world around me.

Math has always been a part of my life; whether it was my dad asking me what angle I needed to shoot my arrow to hit the target, or what the smallest angle I could make on the water was without falling when single skiing. Questions like that created curiosity about the outdoor world and my school math class. That curiosity has only grown with time, and today it is a driving force for how I teach math. I still cannot go anywhere without seeing math, and I love to bring the outside world into my classes as a way of engaging students in mathematics. Technology definitely helps me do that. Technology allows my students to open their mathematical eyes as they learn that math is a part of everything.

When I am in the classroom, my goal is to help students find their curiosity about mathematics by engaging them in their own research and abilities. Technology is a powerful teaching tool for learning.

Introduction

Digital tools allow us to shift how students interact with math. We can use these tools to help our students communicate their ideas, collaborate, and demonstrate critical and creative thinking. Digital tools can also help us shift away from assigning traditional math problems and move toward creating engaging activities that allow students to use the eight standards for mathematical practices. G Suite for Education is one of the best digital tools for helping us get there.

A math class in which students have access to and use Google Apps looks far different from the textbook- and whiteboard-focused classes we experienced as students. The reason? Google Apps allows the opportunity to interact with students in a more meaningful way. Likewise, G Suite empowers students to be creative, critical thinkers who collaborate as they explore and learn.

Our goal with this book is to share ideas that you can use to incorporate technology for teaching mathematics. Although you will learn how to use a few new digital resources, this book is not just about the technology—it is about what technology makes possible in your classroom. Some of what we offer here may be familiar to you. Other ideas may stretch your brain a bit. We encourage you to keep an open mind, take what you see, and consider how you can adapt the concepts for your specific grade level and students.

Be On the Lookout!

 Throughout this book we've included links to resources, including templates, to make it simple to incorporate the G Suite tools into your classroom.

 This book contains screenshots to walk you step-by-step through the many activities we've featured. We've also included Google Tutorials (beginning on page 87) for each of the Google Apps these activities use. If you are new to G Suite or if you're wondering, *How do I do that?*, watch for this icon.

 Wherever you see this icon, you'll find a shortcut, tip, or otherwise helpful hint to make using G Suite even easier in your classroom.

Sneak Peek at a Google Apps Math Activity

"How do we figure out how many dots there are?"

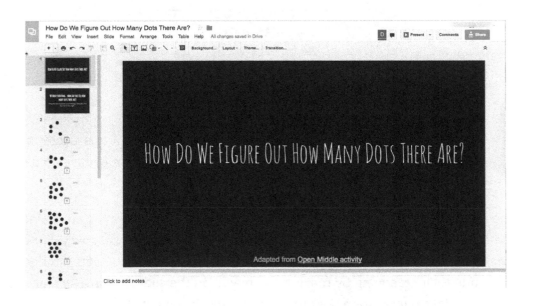

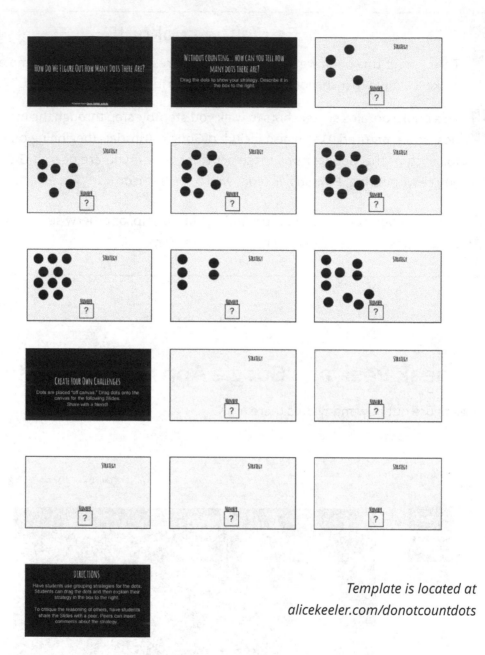

Template is located at
alicekeeler.com/donotcountdots

In this Google Slides activity, students move beyond demonstrating an ability to count. As they interact with the document, they are asked to demonstrate and explain strategies for counting. After devising their grouping strategies for determining the number of dots, students invite peers to critique their reasoning by making comments on the slides. Going further, students create their own dot example for a peer to count through grouping strategies.

Before we get into the math activities, here's a quick overview of how to get going with teaching math with Google Apps in G Suite.

Teaching Math with Google

1 ### Post Directions
Spend more time with students. Post directions in a Google Doc to get them going.

2 ### Watch Students Work
Feedback while students work increases student learning and motivation.

3 ### Collaboration
Google Docs makes real collaboration possible. Give students projects to solve and organize in Google Slides.

4 ### Shift Students to Higher DOK Levels
Focus on thought process over algorithmic steps.

5 ### Students Research
Shift from giving information to asking complex questions.

6 ### Shift to Facilitator
Using Google Docs, Classroom, Search, and YouTube frees the role of the teacher to guide students through their critical thinking.

7 ### Conversations for Deeper Understanding
Instead of grading, engage students in conversations around math.

1. Post Directions

One of the most useful aspects of technology in the classroom is that digital resources can be accessed independently by students. When teachers post directions in Google Classroom, a class website, or on a Google Doc, students can spend more time on task because they spend less time listening to teachers explain assignments. By posting directions, teachers also reduce the number of times they have to repeat themselves. Initially, students are not used to accessing directions and basic content information independently. Expect initial resistance. We do students no favors by doing the thinking for them. An important life skill is being able to figure things out; believe that they can and should be able to figure things out. Directions documents include assignment instructions, as well as instructions for setting up groups or activities. Depth of Knowledge (DOK) 1 or 2 level information can also be provided on a Google Doc to allow students to start working with the information.

 Resource: goo.gl/RMRkdu

2. Watch Students Work

Mentally, when a student is done with an assignment, they are *done*. For this reason, feedback that is provided to students after they have submitted their work does not have the same impact as feedback that is provided while they are working. As students shift to higher levels of critical thinking, Google Docs allows the teacher to get into a document alongside the student to insert feedback and ask questions that help direct students' mathematical thinking. (See Activity 13 for more on initiating feedback conversations.)

 Tip: Use the keyboard shortcut Control+Alt+M to insert comments on student documents.

3. Collaboration

G Suite provides opportunity for unlimited collaboration. Students and teachers can easily collaborate on projects and even work simultaneously in the same Google Doc, Slide, or Sheet. To make the most of this technology, design math activities to be collaborative.

4. Shift Students to Higher DOK Levels

Google Search and apps such as Wolfram and PhotoMath provide answers and steps to DOK 1 and DOK 2 math problems. Ask students to use these tools to enhance their comprehension. Instead of simply looking for the answer, have them focus on the thought process and understanding by approaching the problem from different directions and explaining the solution.

5. Students Research

Complex questions that require multiple searches and the gathering of evidence allow students to more deeply explore their mathematics. Google Sheets is an excellent tool for organizing information, creating graphs, writing equations, and displaying results.

6. Shift to Facilitator

When students are tasked with using technology to do the research, the teacher's role shifts from information giver to *learning facilitator*. In a math class, specifically, the teacher's responsibility shifts from distributing information to creating critical thinkers. By using questioning techniques that take students' understanding to higher DOK levels, you can facilitate creation, exploration, and critical thinking.

7. Conversations for Deeper Understanding

The tools within G Suite's apps enhance a teacher's ability to really communicate with students. Not only do these tools make it easy to offer feedback (including grades), they also open the door for conversations that push students' thinking and understanding. From commenting privately to students in Google Classroom to posting open comments on a Google Doc, G Suite makes it easy to have conversations with students about their work.

"Live in DOK 2 and DOK 3, visit DOK 1 and DOK 4."

—Shelley Burgess, author of *Lead Like a PIRATE*

50 Activities to Teach Math using G Suite

Shifting from using primarily pencil and paper to digital tools in the classroom requires a shift in mindset. This move is not about a quest to go paperless; it is about creating better learning opportunities and deeper understanding for your students. Inserting the same old set of problems and into a digital tool will not make learning meaningful, much less magical. For technology to really be effective in changing the way learning gets done, the task itself needs to become more engaging. And when you combine a better task with powerful yet simple technology, you and your students will see better results—and greater learning. The activities we have included in this book will help you make the shift to technology *and* to improve learning opportunities— one task at a time.

Tip: Start small. Do not begin the shift to a digital classroom by posting *all* of your materials online. A complete overhaul of your lesson plans can be very difficult to manage. Instead, start by using digital tools to interact with students, promote peer collaboration, and create a sense of community in your class.

One simple way to enhance almost any task with technology is to ask the question, "How does this activity make learning better?" As you answer that question you might also ask the following questions:

- Are students more actively engaged in their learning?
- Is data collected to help you respond more quickly to student needs?
- How are students collaborating?
- How does this technology allow for developing critical thinking?
- How are interactions with students improved?

This book is about teaching math with G Suite Google Apps. For each activity, we have identified one or more applications that you can use to implement the idea. We did not design it to be Google Apps training for math teachers, although you will find some Google tutorials at the end of the book.

If you are brand new to G Suite, check out the Google Apps training center:
google.com/edu/training

Find Google Apps tips:
get.google.com/tips/#!/

Sign up for the G Suite for Education newsletter:
goo.gl/H6rkha

This book provides more than 200 resources, examples, and other links. The complete list of links is available digitally at alicekeeler.com/googlemathlinks. The way you use these is up to you. Pick and choose from the activities, or maybe set a challenge for yourself to try one each week. Whatever you decide, taking action is the only way to begin your transition to a digital classroom and teach math with G Suite.

0. Get to Work

As a math teacher, your role is to help students learn and make mathematical connections. One way to elevate your effectiveness in that role is to restrict your words to encouraging students and asking questions that promote critical thinking. Our suggestion: Rather than giving students directions verbally, post them online. Posting directions online allows you to use your time (and your words) in the classroom differently.

When students arrive in class, make their first task locating the directions for getting started. Routine is important; students get into the habit of locating directions and getting to work. Rather than verbally giving or repeating directions, make sure students know where each day's assignments are digitally posted. This simple shift in workflow captures additional time for students to be actively engaged in their learning, since they are not wasting time waiting for instructions.

Alice's Corner

When I switched to posting directions and basic instructions online, I was able to spend more of my time sitting and working *with* students either individually or in small groups. This led to more student *Aha* moments—and that is really what drives my passion for teaching.

Google Classroom

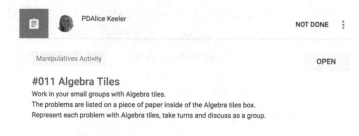

Get started with G Suite by using Google Classroom to post tasks that students are to complete. Students start class by checking Google Classroom online (classroom.google.com) or using the Classroom app on a mobile device to see what they need to get started with for the day. Google Classroom allows the students to mark assignments as *done*.

Google Sites

Google Sites can be used for a classroom website. This can be a place to post the tasks students are asked to do that day. Parents who have access to the Google Sites website are able to ask more specific questions of their child than "how was your day?" Instead, they can say, "I saw you did ___. Tell me about it."

Google Docs

Google Text Docs are web-based, which makes them accessible from any Internet-connected device. Create a Google Doc each day or use the same Google Doc and simply edit daily. The URL to a Google Doc links to the document, not the text in the document. Changing the content of the Google Doc does not change the URL of the document. Share the Google Doc with students and parents so everyone knows what the daily tasks are.

When using a Google Doc to share information, we recommend publishing it to the web. Click on the File menu and choose "Publish to the web." Our suggestion is to create a short URL of the published link using goo.gl or other URL shortening service.

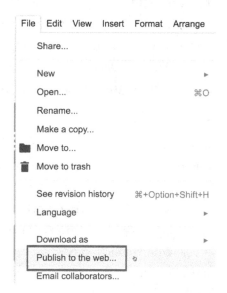

Explore More

A Sample Google Docs Task List that has been Published to the Web:
goo.gl/Z8oQrY

1. Create a Directions Document

Creating directions for an assignment using G Suite apps allows you to share the information in multiple places—with every class and with other educators. Simply provide the link to the directions to anyone who needs access. If you are working with a team of teachers, you can use these tools to collaborate when creating activities. Additionally, you can set the document's permission to allow multiple teachers to update the directions document as necessary.

Google Docs

Providing directions or basic instruction on a Google Doc ensures that every student has the directions. Students who are absent, late to class, or need the directions repeated are able to access the instruction any time. The document can include images, equations, sample problems, and animated GIF's. Images or screenshots can be dragged directly onto the document or pasted in using the "Insert" menu.

No matter the medium, design for student engagement.

A long document full of text is challenging to follow. Break up the document with images, tables, bullet points, and cartoons.

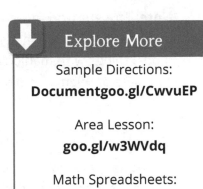

Explore More

Sample Directions:
Documentgoo.gl/CwvuEP

Area Lesson:
goo.gl/w3WVdq

Math Spreadsheets:
goo.gl/H8PQkX

Google Slides

Google Slides are a great way to break information into chunks. Each slide can provide directions, feature an animated GIF or photo example, ask a question, link to a website, show a video, or list steps. Students can click through the slides as they complete each step of the activity, ultimately following the instructions to submit their work.

Google Classroom

Google Classroom allows teachers to assign work and collect it. In Google Drive, create a Google Doc or Google Slides presentation with directions for your students. In addition to text and images, you can attach YouTube videos or videos saved in Google Drive to a Google Classroom assignment. Then, when students log into Google Classroom and locate their tasks for the day, they can watch a short video that models the process, sparks their interest in an overriding question, or prompts them to respond in some way.

 Note: Student work does not have to be digital. The directions document is simply a means of explaining what students are supposed to do. Their task may be to build and model things with manipulatives, to work in small groups to address a problem, or to complete problems from a text.

2. Hear from Everybody

When you pose a question to the class, only a subset of the students will respond aloud. As soon as some students see others start to raise their hands, their thinking may shut down. Technology allows you to hear from everyone. Rather than asking for verbal responses initially, have students respond in a digital format. This gives all of your students time to think and a chance to respond. Then, allow a verbal discussion to build off the digital contributions and ideas of everyone in the class.

Google Classroom

Google Classroom makes it easy to post a digital question. Press the plus icon to use the "Create question" option. Type the question and click "Ask." Within seconds the entire class can respond to the question.

When possible, select the "Students can reply to each other" option. This allows the students to hear from each other, which contributes to the class operating as a community of learners.

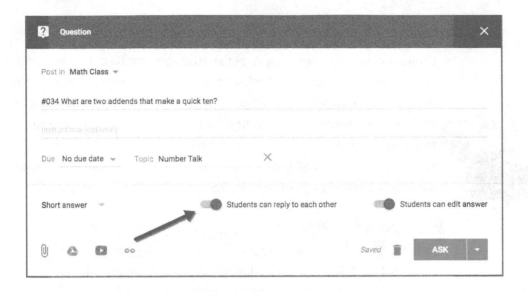

Google Slides

When used collaboratively, Google Slides can facilitate discussion.

 First, create slides with discussion prompts and share the slides with edit access for the students. Then, have students insert a response slide after the prompt. As the slides are added, students can use the commenting feature to respond to and comment on each other's slides.

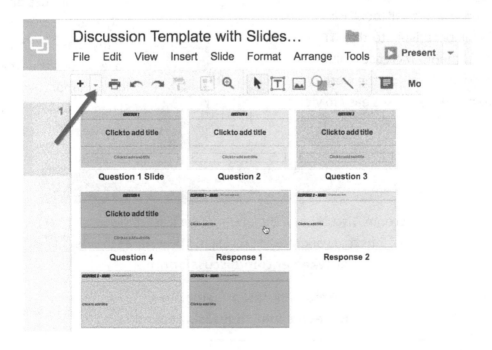

Diana's Corner

Brainstorming is a great way to get all students involved. I have found that giving students a numerical value to reach forces them to get beyond the obvious. For example: Instead of saying, "list everything you know about $y = 3x + 7$," ask them to list 15 things about $y = 3x + 7$. Giving them the option to know nothing is not an option. I have them first brainstorm individually, then work in small groups, then work with the whole class. Whether we're using technology or not, the conversations are empowering.

Explore More

Slides Template—prompt a question and allow all students to individually insert a slide to respond:
alicekeeler.com/slidesdiscussion

Basic Ask and Respond Template:
alicekeeler.com/slidesaskrespond

Google Sheets

In a Google Sheet, each student can type their response into a cell (box). Have students respond collaboratively to the same Google Sheets document. Increase column widths as necessary and turn on the word-wrapping feature to make responses easy to view.

Alice's Corner

Starting discussions digitally has become a staple in my class. I frequently have students tell me the collaborative Google Slides or Sheets was their favorite part of the lesson.

Explore More

Classroom Discussion Template—create a list of discussion questions and generate a discussion sheet for each one:
alicekeeler.com/discussiontab

Group Brainstorm Template:
alicekeeler.com/slidesaskrespond

3. Private Comments are the Assignment

No matter how nice you are, the authoritarian relationship that naturally exists in the classroom may prevent some students from speaking up. It can be really intimidating to ask the teacher a question! Having a digital space to ask a question privately has been shown to increase the likelihood that students will ask questions. When you use the private comments space for assignments in Google Classroom, students will realize they can use the feature to interact with you.

Alice's Corner

While teaching a class, I went around and looked each student in the eye and nicely asked them, "Can I help you? Do you have any questions?" Each one said, "No, I'm good." When reviewing their work, it was easy to see they were not good. It can be hard for students to ask a question when peers are nearby.

Google Classroom

On every assignment in Google Classroom, students can leave a private comment. Create a culture in the classroom that encourages feedback from students by asking questions in private comments. Invite students to go to Google Classroom and reflect on the task in the private comments section. Make a point to reply to each student.

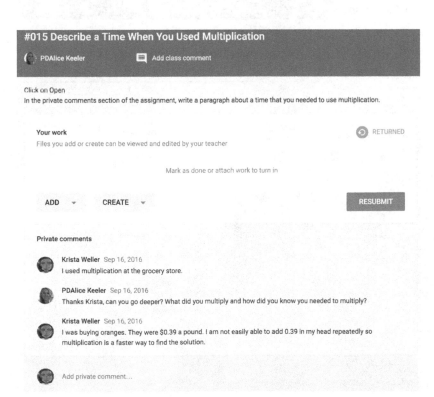

Rather than opening 30–200 individual student documents, use the private comment space for assignments in Google Classroom. This creates an easy way for students to share their mathematical thinking and a quick way for you to view and respond to their comments.

4. Cut Quarter Sheets of Paper

Being paperless, in and of itself, is not a virtue. Remember that our goal is to use technology where technology enhances learning. While it makes sense to store resources, directions, and information digitally, working things out on paper can be the most efficient way to interact with the math..

Tip: Switch from full sheets of paper to quarter sheets of paper. Stock stacks of quarter sheets of paper around the room so that it is easy for students to grab a sheet and jot down a sketch or work out a math concept. The smaller paper encourages students to focus on one math problem at a time.

Google Slides

Google Slides are perfect for capturing images from quarter sheets of paper. Students can hold the paper up to the webcam or use a mobile app to insert the image onto the slide. This allows the student to submit their work easily to Google Classroom by adding the Google Slides. Quarter sheets also ensure there is not too much content on the paper so it can be easily viewed in Google Slides. An entire sheet of paper can be challenging to read on a slide.

Another advantage to adding images of the quarter sheets of paper to Slides is the ability to add feedback conversations. Students can leave feedback for each other by commenting on the Slides.

Explore More

Quarter Sheets Paper Collaborative Template:
alicekeeler.com/quartersheetspaper

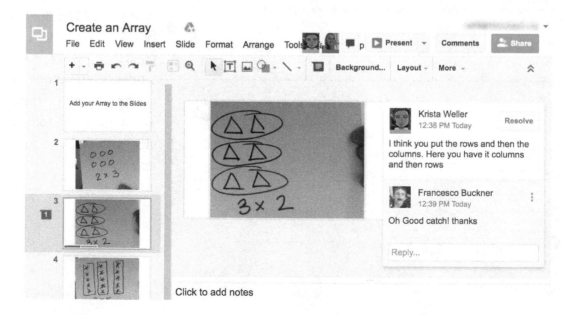

5. Have Students Show Their Thinking

Instead of telling students to show their steps, ask students to "show your thinking." Several websites and mobile apps show students' steps on solving a math problem. This technology is not going to go away and it is only going to get better. If they only have to show the steps used to solve the problem, they can easily find the solution on one of these sites and copy the process.

Do not fight technology. Embrace it and ask, "What can we do now?"

If, on the other hand, they have to show their thinking, the focus of their response shifts from just finding the answer to explaining the reasoning for the solution. One way to have students show their thinking is to ask them to compare and contrast processes to get to solutions for two or three questions/tasks.

Diana's Corner

Ever have one of those days when student work seems too good to be true, but you wanted it to be true? I had one of those mind-shattering days because of one student. One of my students suddenly started producing incredible homework that showed every single step; it was great to see. But when it came time to take the test, the student was unable to replicate that great work and instead was confused and lost. When confronted with the discrepancy, the student responded with "Mrs. Herrington, have you ever heard of Wolfram Alpha?" From that moment on, I knew that I had to change. In today's world, banning technology is not an option. I needed to figure out how to embrace it. My solution was to change how I ask questions.

THINKING...

Google Slides

Google Slides is essentially blank pieces of paper for students to use to show their thinking. Rather than an endless piece of paper, Google Slides allow students to organize their thoughts in a limited space. An expectation should be that students need to clearly communicate their ideas. Aside from its usefulness for math class, using Google Slides teaches students to create attractive presentations that clearly communicate what they are thinking, and those are both great skills for students to master.

Students are able to critique the reasoning of their peers by sharing the slides. The blue "Share" button allows students to add the email addresses of peers to allow for peer editing. Students can allow comment-only access or edit access. In the event that one student accidentally "messes up" another student's work, utilize the revision history to restore the correct version.

Explore More

Add a Blank Slide Template:
goo.gl/6VAjTD

Add a "Description of Your Thinking" Template:
goo.gl/wjT75h

Ronessa Acquesta Operations Card Sort in Desmos:
goo.gl/z9qGwX

Google Doc

Creating a table in Google Docs constrains the size of images in the table. Insert screenshots from Google Search, Wolfram Alpha, or other applications that show steps to a math problem into a table to show the solutions side by side.

Diana's Corner

Asking students to look at worked-out solutions and discuss how the processes were the same and where they were different opens up mathematical understanding. Surprise yourself! Try this as the introduction to a new process and see how quickly students jump in on the conversation.

Explore More

Compare Solutions Template:
goo.gl/iBVGms

Show Your Thinking Example:
goo.gl/ItwQL2

Photomath App:
Android **goo.gl/RcqwiD**
iOS **goo.gl/5vs4DO**

6. You Want to Eat a Brownie

Sometimes math problems provide too neat of a situation where all of the necessary information is offered to the student. Nothing more, nothing less. To increase critical thinking and allow for student creativity, do not provide the students with all of the information. Have the student ponder over what information they need to know.

In this "You Want to Eat a Brownie" activity, students are asked to think about what they need to know in order to answer the question, "How many minutes will you need to use the exercise bike to balance the number of calories from your brownie?" We recommend having students work in small groups to brainstorm a list of things they need to know.

This problem can be approached in a variety of ways and can be adapted to the students' level of math. As students develop their list, ask them questions about their questions. "When I see a plate of brownies at a potluck, I notice that someone always cuts one in half and leaves the other half . . . how would that influence this outcome?" Students may surprise you with what things they come up with that add to the complexity of the problem. Students may suggest, "What is the tension level of the exercise bike?"

To answer their own questions students will need to identify assumptions and look up information. For example, an assumption students need to make is the size of the brownie. They will need to look up a brownie recipe to determine calorie levels and then adjust for the size of their brownie.

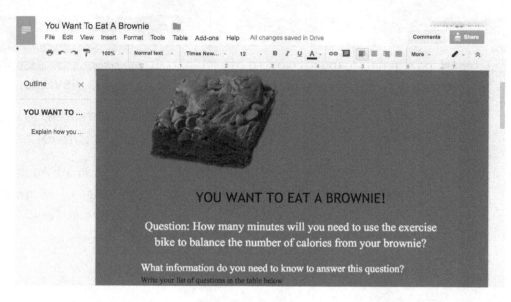

The purpose of this activity is not to necessarily get a correct answer, but rather to help students develop their thinking and to have them clearly communicate their ideas about this problem. You can also ask students to create an infographic to explain their strategy and any generalizations they notice. This helps students focus on how they tackled the problem and aids in evaluating student thinking.

To bring in an authentic audience, students could make brownies of different sizes and offer them to people. Students can record the data they need in order to tell the person how many minutes they need to spend at the gym. If students determined a formula prior to offering the brownies and had it in a spreadsheet they could offer the information on the spot. Otherwise, collecting the person's email address would allow the students to create a professional report and send the information at a later time.

Explore More

"You Want to Eat a Brownie" Sample:
alicekeeler.com/eatabrownie

7. Look up Information

Teach like YouTube and Google exist.

When designing lessons, ask the question, "What are students looking up?" Instead of giving students information, ask them for information. Rather than telling them, "The quadratic formula is . . ." ask, "What is the quadratic formula?"

Google allows people to do a lot more than search the web. Google allows students to ask questions about converting measurements, graphing parabolas, and a lot more. In today's culture, people use a search to look up anything they wonder about. Searching for information occurs daily in social and work situations. So, why not allow students to use Google? Recognize what a Google search can reveal and shift your assignments to include the types of questions and tasks that students are already doing.

What questions can students ask Google? Design activities that teach students how to effectively use Google calculators and search for answers. Showing students what they can do with Google provides them with the valuable life skill of knowing how to find information. Then, what the students do with the information becomes the focus of deeper learning.

Today's students (and adults) have access to apps that do all the number crunching for them. Instead of pretending this technology doesn't exist, teach students how to look up, contextualize, and use the information that's at their fingertips.

Alice's Corner

I have a rule: Do not tell students things they can look up.

Google Search

On a mobile device or within the Chrome browser students can say, "Ok Google" to verbally ask a question. Alternatively, students can type in their question. Try the questions/problems below:

- What is the area of a rectangle with length 10 and width 5?
- What is the circumference of a circle with a diameter of 10?
- What is the definition of circumference?
- How do you find the volume of a cylinder?
- What is y equals negative 3x squared minus 8x plus 2?
- Perimeter of a trapezoid.
- What is the square root of 56?
- sqrt(12)
- How many feet are in 23 meters?
- Volume of a cylinder with radius 10 meters and a height of 5 cm.

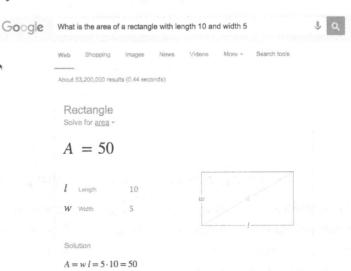

- How do you find the radius of a sphere given the volume?
- Triangular prism calculator.
- Calculate the diagonal of a cube with 12.2cm edge.
- a^2+b^2=c^2 calc a=9 b=3 c=?
- z=3x^2-4y^3
- Graph multiple graphs together: (1/3)x/2, sin(ln(x)/3), -3x^2, sqrt(x)

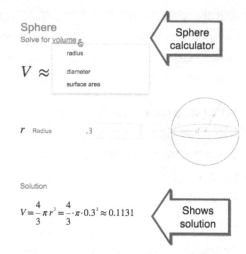

Explore More

Google Slides Activity Template to "Ask Google": **goo.gl/ODtTf3**

Try the Wolfram Alpha Add-on for Google Docs. Bring additional search and calculator functionality to Google Text Documents: **goo.gl/ixtIA3**

Ok Google Voice and Search Actions: **goo.gl/TQLcgb**

Google Calculator and Unit Converter: **goo.gl/QNys4a**

Google Search Tips and Tricks: **goo.gl/oDMNud**

8. Create a Blank Document

The person doing the work is the person doing the learning.

While templates for students can help guide a lesson, they are not always necessary. Just as we would ask students to "get out a blank piece of paper," students can create a blank Google Doc on which to work out and work through problems. We believe that Google Docs is an improvement over paper for the following reasons:

- It allows more collaboration opportunities.
- Student and teacher always have access to the document, both before and after the due date.
- There is room for revisions and mastery learning.
- It allows commenting and feedback opportunities.
- It encourages feedback *conversations.*
- It makes peer evaluation possible (and easy).
- It provides access to the *Explore* pane for research.

Even if they are using non-Google tools, students can insert screenshots of what they have been working on into a Google Slide or Doc. Take the learning one step further and have students explain what is in the screenshot. By organizing information in Google Slide or Google Docs, students can easily communicate what work was completed and their thinking around it.

Alice's Corner

When demonstrating math problems on the board or on paper, I consistently end the problem with how it would look in a spreadsheet. For example =sqrt(7)/(x + 7). I asked that students, when using paper, were consistent about this also. When typing math into a Google Doc or responding in a Form, students could also easily adjust to entering spreadsheet notation.

"Type Your Math" Sample: goo.gl/V9y2cu

Google Classroom

In Google Classroom, students click on the "Open" button in an assignment to view the Student Work section. Students can add documents from their Google Drive or from their device, or link to their work if it is a website. The "Create" button in the assignment allows students to create Google Docs, Sheets, Slides, or Drawings right in Google Classroom. This gives the teacher instant access to the students' work, regardless of whether the blue "Turn in" button is selected.

This process also automatically adds the student's name to the document title. (No more nameless papers!)

Google Drive

If Google Classroom is not being used, documents can be created in Google Drive. Files created in Google Drive are, by default, private. The student will need to share the file with the teacher by selecting the blue "Share" button. Having students create and share a folder in Google Drive with the teacher makes it easy for the teacher to access student work, as anything in the shared folder is automatically shared with the teacher.

Google Slides

Blank Google Slides make an excellent medium for students to demonstrate their mathematical processes. Students can do one problem per slide or show each step in a process. The ease of organizing text boxes, images, videos, or drawings onto a slide make this a wonderful tool for clearly communicating ideas.

Explore More

Doctopus copies templates of documents for students and allows the teacher to share and monitor docs. A folder structure for each student is created in Google Drive:

goo.gl/PzCV9r

Template to Create a Folder for Each Student:

alicekeeler.com/makefolders

Diana's Corner

Google Slides allows students to slow down and focus on their thinking. Using a slide requires thought and organization.

9. Create Community

We are a community of learners and we help each other get better.

The time spent in building culture and community in your classroom is an investment. In fact, the most important thing we do as teachers is build relationships with our students. One way to get to know the students and begin to familiarize them with Google Apps is to have them introduce themselves on a collaborative Google Slides presentation. Have each student add a slide, insert a selfie, and share a little about him or herself. Since all student slides are in the same presentation, pressing the "Present" key makes it easy to greet everyone. This also provides a study sheet so you can quickly memorize your students' names.

Explore More

Collaborative Introduction Google Slides Template:
alicekeeler.com/meetmemath

10. View Form Responses in a Spreadsheet

Google Forms is an excellent tool to gather information and data. Information submitted through a Google Form can be saved in a single Google Sheets spreadsheet. This gives the teacher one place to view student submissions. Teachers can view student submissions instantly, which means the data can be reviewed and used to make immediate adjustments to instruction.

After creating questions in the Google Form, the "Responses" tab enables the creation of a spreadsheet to go with the Google Form. Form responses can be

viewed in summary form on the "Responses" tab or within the spreadsheet. To enable the spreadsheet, the "Create spreadsheet" icon must be clicked. The spreadsheet can be created before student responses, while students are responding, or after students respond. The spreadsheet will, in all instances, display all of the Form responses.

Multiple choice questions are great for formative assessment since they allow for a quick response to the data. Google Forms automatically creates summary charts of responses. Assuming a spreadsheet has been created, deleting responses from the Form does not remove responses from the spreadsheet. This feature allows you to delete responses after each class to allow for a refreshed summary of responses.

Go deeper with student data by clicking on the spreadsheet icon on the "Responses" tab in the Google Form. All student responses are in a single spreadsheet. One option for reviewing student responses is to sort each question column. This allows you to see, question by question, which specific students are struggling with a particular concept.

💡 **Tip** To use the same Google Form and spreadsheet for all classes, make sure the form includes a question about what period or section the student is in. This allows you to sort and filter based on your classes.

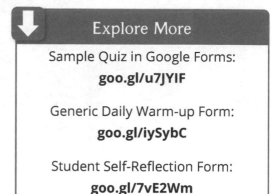

Explore More

Sample Quiz in Google Forms:
goo.gl/u7JYIF

Generic Daily Warm-up Form:
goo.gl/iySybC

Student Self-Reflection Form:
goo.gl/7vE2Wm

11. Use Pixel Art

Pixel art is an excellent tool that students can use to model their mathematics or explore a concept. Pixel art can be done on graph paper or in a grid on a spreadsheet. In Google Sheets, students "paint by numbers" by entering in single digit numbers that correspond to a particular color. Pixel art activities can be used from Kindergarten all the way through to Calculus-level, where the individual blocks may represent any number of concepts including integers, division rules, place value, and polynomials.

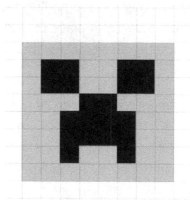

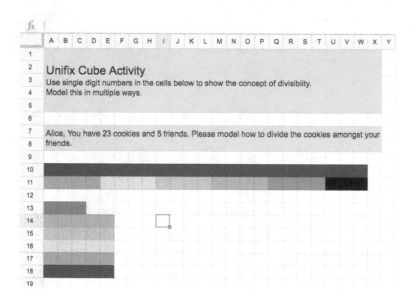

Alice's Corner

Students love pixel art! Be on the look-out for smiles.

Google Sheets

Adjust the height and width of cells in a Google Sheets spreadsheet to create a pixel art grid. Using conditional formatting on the cells, students can type in a value and have the cell color fill. The trick is to make the font color the same as the fill color in the conditional formatting rules. Create conditional formatting by selecting the range of cells and use the Format menu to choose "Conditional formatting." A side panel will appear allowing you to define the conditions for formatting.

You can use a grid in Google Sheets in several other ways. For example, students can use the grid to measure items in units. Placing pictures on the spreadsheet, students can count and measure utilizing the pixel art.

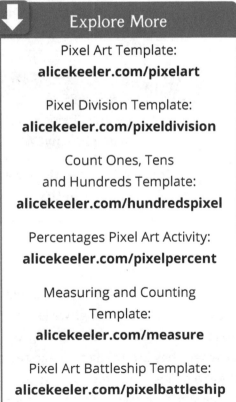

In a pixel art spreadsheet, students can also create their own picture and estimate the percentages of the colors they used. After estimating, the student can use cell referencing to calculate each percentage and then explain how the results make sense.

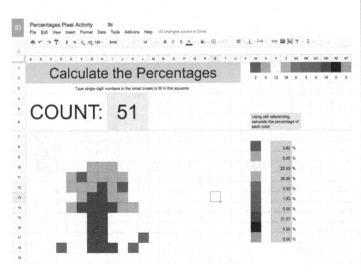

Explore More

Pixel Art Template:
alicekeeler.com/pixelart

Pixel Division Template:
alicekeeler.com/pixeldivision

Count Ones, Tens
and Hundreds Template:
alicekeeler.com/hundredspixel

Percentages Pixel Art Activity:
alicekeeler.com/pixelpercent

Measuring and Counting
Template:
alicekeeler.com/measure

Pixel Art Battleship Template:
alicekeeler.com/pixelbattleship

12. Guide Instruction

Traditionally, instructional time is used for the teacher to present definitions, steps, and basic examples— and then waiting for students to copy that information. Rather than having students focus on simply copying notes, teachers can empower students to take ownership of their learning by providing them with guidelines and encouragement.

Google Forms

Google Forms supports a flipped model for teaching. Insert an image or YouTube video directly into the Form for students to respond to. Add comprehension questions following the video or text to ensure students actually watched the video and/or read the text. When students submit the Google Form, you will see a record of who went through the lesson and which students are struggling with the material.

Diana's Corner

Guiding learning allows students to be creative and engage in their learning more. You have become their academic coach, and you fine tune their learning.

Concentric Circles

Play the short YouTube video below to hear the pronunciation of Concentric Circles. Explore concentric circles and respond to the questions in the Google Form.

Concentric Circles Pronounced

Concentric Circles Definition

Concentric circles are **circles** with a common center. The region between two **concentric circles** of different radii is called an annulus. Any two **circles** can be made **concentric** by inversion by picking the inversion center as one of the limiting points.

> **Explore More**
>
> Concentric Circles Activity:
> **goo.gl/YcmyLW**
>
> Concentric Circles Google Form:
> **goo.gl/upULgN**

13. Feedback Conversations

We've mentioned before that Google Apps allows students and teachers to collaborate on learning. One of the things we like most about the collaborative features of G Suites Apps is the ability to comment on a work in progress. When you make comments as students are working or when they are ready to make revisions, you can help them think more deeply about what they're doing. Initial feedback to a student is the beginning of a conversation to help students understand a process and move forward in their learning.

Feedback conversations can also be among peers, where they help one another edit and revise their work. With forty students in a class, peer conversations are a "win-win" for everyone.

The keyboard shortcut for comments, Control+Alt+M, is a quick way to communicate. Comments can also be added through the Insert menu. Students (and teachers) are able to reply to each other's comments, allowing for conversations around the task.

Comments help to facilitate asynchronous collaboration. One student is able to make observations about the math in the document. At another point in time, a collaborating student is able to respond to those comments. Feedback conversations that develop in comments—both from peers and from you as the teacher—are a valuable way to help students to develop their critical thinking around mathematical concepts. And when comments are offered before students are mentally "done" with a task, you increase the value of the comment since students are more likely to read, consider, and revise their work if necessary.

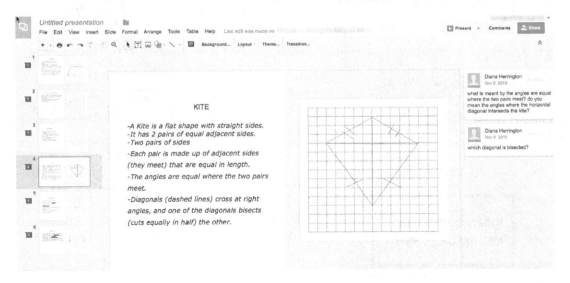

Google Docs

Highlighting a word, phrase or element in a Google Doc allows for inserting a comment on that selection. A button appears in the right edge of the Google Doc that inserts a comment note in the margin. Alternatively, use the keyboard shortcut Control+Alt+M to insert the comment and Control+Enter to save the comment.

Google Sheets

Google Sheets makes for a great collaboration tool. While comments in a spreadsheet are not left in the margin of the document, they can be easily found under the "Comments" button in the upper right-hand corner.

Explore More

Create a spreadsheet per student to receive ongoing feedback:
alicekeeler.com/commentmaker

Create a list of feedback per student and email it to them:
alicekeeler.com/sendcomments

Alice's Corner

Leaving handwritten comments on a student's paper does not invite conversations. Moving from comments to conversations is truly one way that Google Apps transforms my classroom.

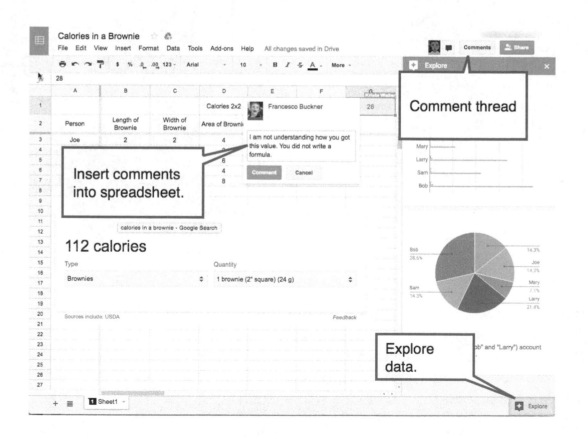

Google Classroom

Work assigned in Google Classroom provides the teacher immediate access to student work, even before the student has started the work. Student work can be located by clicking on the assignment title in the Stream. Student names are listed on the left-hand side of the stream. Clicking on any student's name shows work they've been assigned. To open a student's work, simply click on the name of the work to open the document. When using Google Docs, Sheets, Slides, or Drawings, feedback comments can be inserted right in the document at any point of the process. Be aware that, with Google Classroom, when students turn in work they are removed as the owner of the document and made a viewer. Viewers do not have access to comments left in a document. Teachers must return the student work in Google Classroom for students to see in-document comments.

For each assignment, there is a private comment option for each student. Students can ask private comments or respond to private comments left by the teacher. These comments in Google Classroom are always visible to students.

Regardless of where comments are made, we encourage you to view them not as static comments but as conversations.

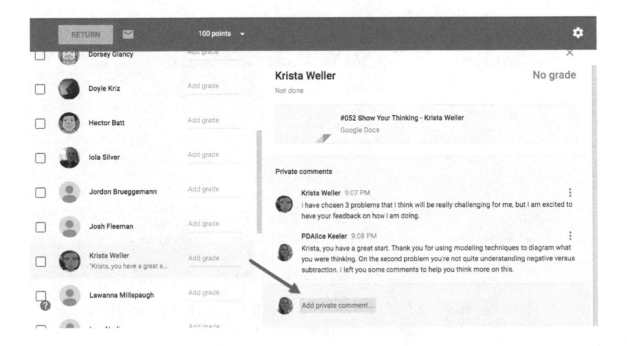

14. View Student Work in Google Drive

Rather than managing a folder or box of papers, digital work stored on Google Drive can be easily accessed from any Internet-connected device. Simply install the Google Drive, Docs, Sheets, Slides and Classroom apps on your computer or mobile device.

Google Drive

Unlike paper, files in Google Drive provide data about student and teacher
access. "Recent" on the left-hand side
reveals files that have been recently
edited; this allows for fast response to
students working on documents.

Sort a folder of student work by "Last opened by me." The date you last looked at a student doc is listed. If you have never looked at the document it shows a minus sign. This signals to the teacher that this assignment needs feedback.

Explore More

Drive20 Chrome Extension:
alicekeeler.com/drive20

Students can submit a folder of work in Google Drive with this Sheets Add-on:
alicekeeler.com/turninfolder

List student assignment work in a spreadsheet to keep track of providing feedback:
alicekeeler.com/gradethefolder

Google Classroom

A paradigm shift for using Google Classroom is that "attachments" are not attachments; they are links to content that is stored in Google Drive. Clicking on the title of any "attachment" launches the document from Google Drive.

For each assignment, a folder is created for the teacher in Google Drive. This keeps all student work neatly organized in Google Drive by assignment. The assignment folder is easily accessible on the assignment assessment page. Click on "Student Work" along the top to view the number of students who are "Done" or "Not Done" with the assignment. Underneath the count is a folder icon. Clicking on this icon opens Google Drive to reveal all student work.

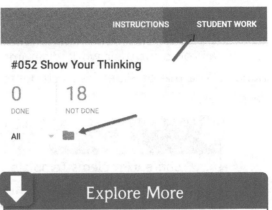

Explore More

ListDocs Add-on for Google Sheets provides a list of all work that has been submitted to Google Classroom: **alicekeeler.com/listdocs**

15. Choose 3 Problems

"Practicing an isolated skill reduces the DOK level. Students do not have to think about what steps to apply."
—Shelley Burgess

High-quality feedback improves learning, whereas grades can detract from learning. To boost students to higher critical-thinking levels, we must give them the opportunity to think, receive feedback, and respond to that feedback.

Assignment	✕

Post in Math Class ▾

#052 Show Your Thinking

Choose 3 math problems from the text.
Create a Google Slides in Google Classroom by clicking on OPEN and clicking on CREATE.
Go beyond procedural steps. Model each problem and demonstrate understanding.
During this process I will be providing feedback. Inserting questions to help develop your thinking.
Respond to the comments and update your work.
You are NOT graded on having the correct answer.
Instead you are assessed on how you respond to the feedback comments.

Due No due date ▾   Topic Show Thinking   ✕

Saved   **ASSIGN** ▾

Google Slides

Instead of having students do a pre-determined, repeated set of problems, consider having them choose three problems. Students show their thinking, diagram and model the questions, and go deeper in their learning. Students do their work in Google Slides, which has room for their work, diagrams and explanations. The use of Slides helps students focus their responses.

Explore More

Share 3 Problems Template:
alicekeeler.com/3problems

Diana's Corner

At first, my students chose the easy problems, but when they realized that I was really going to help them develop their thinking for their chosen problems they began to choose more difficult problems. They felt free to be risk-takers.

16. Conversations Are the Assignment

"Failing is part of the learning process."
—Dr. Jo Boaler

Oftentimes, students perceive math as being about getting the correct answer. With this in mind, make the conversation about a few problems in the assignment. Instead of asking students to "do all odd numbered problems," allow students to choose three to five problems that they would like to solve. Students are to do no more than one problem per slide. Their assignment begins with doing the problems, but the valued part is to carry on a conversation through comments from their teacher or peers about their solution process. When it comes to math, learning happens when you talk about it and question the process. As comment conversations become a regular part of learning in your classroom, students will discover that it is *during* the conversation that they grow in their abilities. Equally as important, they know that the person (or people) with whom they are carrying on the conversation cares more about their thought process than the answer. The conversation becomes a risk-free learning zone—and that's where the magic happens!

Google Slides

Google Slides is a strong platform for communicating with students. It allows for a student and teacher to have one-on-one time at any given time. Students can use one slide per step to demonstrate the problem-solving process in great detail. Using either comments or the "add notes" section, you (or perhaps a classmate) can ask guiding questions to help the student understand what they have done and possibly offer direction for improving upon or expanding their work. Students have the opportunity to ask questions and/or edit what they have done. This kind of interactive instruction and problem-solving through selected problems or tasks allows students to acquire a stronger understanding and feel more confident with their learning.

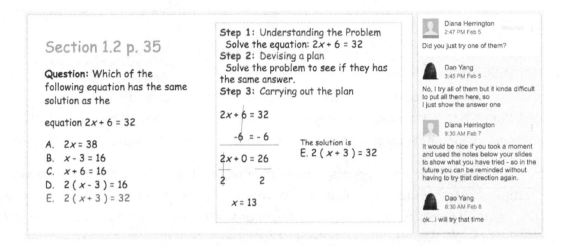

17. Bitmoji Feedback

Use the Bitmoji Chrome extension to insert fun feedback for students in the form of a graphical avatar. Adding the Chrome extension places an icon to the right of the Omnibox (address bar). Click on the Bitmoji Chrome extension to choose an avatar that expresses the intended feedback. Drag the Bitmoji avatar directly onto a Google Doc, Google Slide or Google Drawing. Note that this does not work with Sheets or Forms.

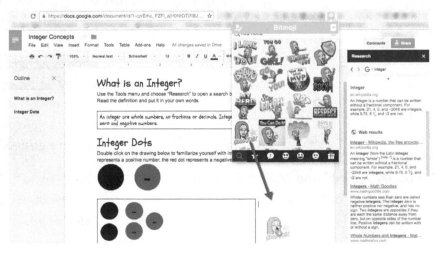

Google Chrome

The Bitmoji Chrome extension can be added through the Chrome Webstore. Create an avatar that fairly closely represents you. Putting "yourself" into digital work is shown to help students to feel more connected to the teacher and the class.

Explore More

Bitmoji Chrome Extension:
goo.gl/aUJnf7

18. Have Students Put Themselves in It

When all students submit the same work, they are not being creative—and they probably aren't thinking critically about their work. When designing tasks, consider this question: What decisions can each student make that may be different from their peers? The point of this exercise is to think about what you can do to allow for some individuality within the assignment.

Construct activities that invite students to use something of their choosing. It could be as simple as letting students choose the objects with which they represent their math. When students can use their own pictures (or pictures they choose off of the web), they are able to insert a little of their personality into the assignment.

Alice's Corner

When creating tasks, I try to avoid having thirty of the same thing submitted. This also helps to reduce copying if students have to include some original elements.

Google Slides

Students can personalize Slides by adding their own images to a slide. Images can be dragged onto the slides, added from a mobile device, or inserted from images found using the "Search" tool. By default, the search is for all image types. This can be filtered for "Clip art" or "Photo" depending on what the student is looking for.

Explore More

Create Probability Trees Activity:
goo.gl/keWl8H

Constructions Activity:
alicekeeler.com/slidesconstructions

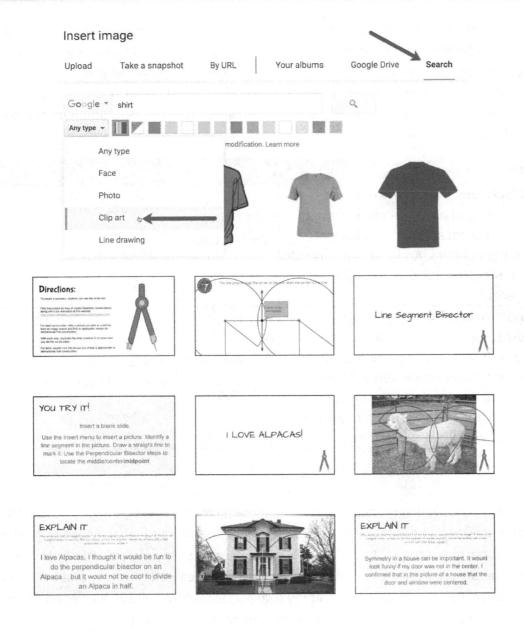

19. Use Slides for Learning

Discovery is an important part of learning. One way to get students engaged in learning is to have them share their discovery or investigation process. Ask students to research a topic, such as "the properties of diagonals and interior angles of a quadrilateral." Then have students create a Google Slides presentation to demonstrate their step-by-step investigation, making sure to include diagrams, photos, and their findings. Slides allows students to demonstrate their process rather than simply the final answer. The learning changes from "getting an answer" to "showcasing the process."

Throughout the process, encourage students to collaborate with you and with each other by carrying on conversations using the Comments tool in Slides. Providing comments to students *during* the investigative process increases student motivation and understanding. This collaboration and feedback should become part of the learning process and can be part of the grading process, as well. For example, make student comments a requirement (and part of the grade) on portfolios and projects.

Topic	Point Value	Point Value	Total Score
Project: Two projects are required per semester, and must include a comment/feedback from at least two different members of class. And you must provide a comment/feedback to two different projects.	Written Project Value 30	Feedback you give to someone's project 10 each	50 points
Portfolio: Semester long portfolio addressing your CCSS progress with cover letters for each entry. Must include a comment/feedback from at least two different members of class. And you must provide a comment/feedback to two different portfolios.	Portfolio Value 100	Feedback you give to someone's portfolio 50 each	200 points

Diana's Corner

In a math investigation of the Fibonacci sequence, I created a Google Slides document for my students. Wanting to make sure they understood the activity, I felt the need to give them information and examples to look at before they began. So my classes followed my examples. After stepping back and reflecting on my activity, I realized creativity and individuality were not really there; I was looking at my own presentation over and over again. I needed to figure out a way to open it up so they could be seen in their work.

Looking differently at technology was my answer. I didn't give examples, but instead had students use the Explore tool to find the examples. Conversation during the investigation changed, students shared examples they were finding, and they were amazed at how many things use Fibonacci numbers. They looked for patterns beyond what students had done in the past, and they also reflected their individuality and creativity. Ultimately, their work was stronger and they developed higher levels of understanding than if they had simply followed my example.

Explore More

Original Fibonacci Exploration Activity:	Calculus M&M Activity—Tech Enhanced:
goo.gl/vPhbN3	**goo.gl/TODsRw**
Revised Fibonacci Exploration Activity:	Investigating Sine and Cosine Example:
goo.gl/Q2nAEi	**goo.gl/0p3mD6**

20. Small Group Investigation

Rather than starting a class session by having students take notes, allow students to investigate a concept in small groups and collaborate on the same Google Slides template.

Group members add their name to the document title. Students add additional slides in order to insert their responses and evidence.

Google Slides provides the opportunity for students to do research directly in the document. Under the Tools menu, students choose "Explore." This opens up a side panel that allows students to complete a Google search. Rather than telling students basic information and formulas, ask questions. "What is the quadratic formula?" Students are able to add the answer to the slide by dragging it from the Explore pane, or by entering it onto the Slide. Once students have that information, the question becomes *what do they do with that information?*

Diana's Corner

I am always surprised with my students' group investigations. Technology has ramped up the quality and the DOK levels with their ability to collaborate. A win-win learning tool.

If students are working in small groups of four, have each student in the group be responsible for investigating one trait of a concept. After investigating individually, students who studied the same traits temporarily group up together to compare notes. Students then return to their original group to share what they know.

 Tip Desmos (desmos.com) is an excellent tool for allowing students to move past procedural steps and to focus on recognizing patterns.

Going further, groups of students are asked to discuss and analyze an application of what they just practiced.

Looking with the parent general graph
$ax + by + c = 0$

An equation for a line can be written many different ways. We are going to look at the general form of the line $ax + by + c = 0$

Investigate what changing the a, b or the c does to the graph of the line. Make sure to have your own line on the same graph that you are plotting.

In your group you are going to each investigate one trait. To investigate the trait choose values for a, b and c, and share what you learn with your group:

- Trait 1: how does changing only "a" transform your original graph?
- Trait 2: how does changing only "b" transform your original graph?
- Trait 3: how does changing only "c" transform your original graph?

Transforming the parent function (graphing)

$y = \cos(x)$ is the parent function for cosine. There are several things we can do to alter its graph -- we are going investigate how each changes the graph.

General formula for a cosine curve is: $y = a \cos(bx + c) + d$

In your group you are going to each investigate one trait, and share what you learn with your group:

- Trait 1: $y = a \cos(x)$ how does a transform the graph?
- Trait 2: $y = \cos(bx)$ how does b transform the graph?
- Trait 3: $y = \cos(x + c)$ how does c transform the graph?
- Trait 4: $y = \cos(x) + d$ how does d transform the graph?

Looking at the parent graphs

When graphing sine and cosine curves you need to look at the parent graphs and how you can transform them. *Note: when using a graphing utility make sure you are in radian mode.*

Parent Graph for sine: $y = \sin(x)$ Parent Graph for cosine: $y = \cos(x)$

Graph $y = \sin(x)$

- What do you notice about the graph -- what are the characteristics?
  - Minimums? Maximums? Increasing? Decreasing? Zeros? Repeating patterns? Change in curvature?

Google Slides

Students can collaborate in small groups on a single Google Slides investigation. Often, when students are investigating concepts and mathematical traits in class, they come across real-world examples of those traits after school or at home. The Slides app makes it possible to quickly add such examples to the group project.

Google Docs

Google Docs can be an excellent collaboration tool for small groups. One way to ensure individual participation in the group project is to add a table that provides a space for each student to include their response in the Doc.

Structure the activity so the Google Doc begins with a question that students can discuss, research, or investigate.

Explore More

Sine and Cosine Investigation:
goo.gl/t4JBxu

Linear Equations Small Group Investigation:
goo.gl/wNIORA

Rounding Whole Digit Numbers Small Group Investigation:
goo.gl/Jfmcb0

 Tip To create answer lines in a Google Doc, create a 1x1 or a 2x1 table. Hold down the SHIFT key and select the edges of the box, except for the bottom edge. In the toolbar, select a line weight of zero.

Slope Intercept Form

Small Group Members

Research
Use the Tools menu and choose "Explore." What can your group find out about slope intercept form? Put your results below....

Contributor	Information or Picture

The Formula
Find the formula for slope intercept form and put it in the box below

Use Desmos
Type the formula for slope intercept form into a new desmos.com activity. Use the sliders to investigate what the variables do. Insert Screenshots into the table cells below

From your investigation, what did you discover?

Explore More

Small Group Investigation Google Docs Template:
goo.gl/2ylmxx

Directions for Creating Answer Lines in a Google Doc:
goo.gl/En4rEk

Rounding Whole Digit Numbers Small Group Investigation:
goo.gl/Jfmcb0

Google Sheets

Creating multiple sheets in Google Sheets allows students to think and write about one concept at a time.

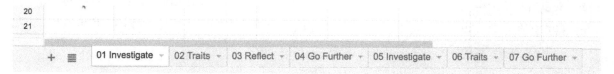

 Tip: Longer response boxes in Google Sheet can be created by merging cells and setting the word wrap on the merged cell. Use the paint can tool to fill in the merged cell. This helps direct students to know where to fill in their response.

Explore More

Small Group Investigation
Google Docs Template:
alicekeeler.com/sheetsinvint

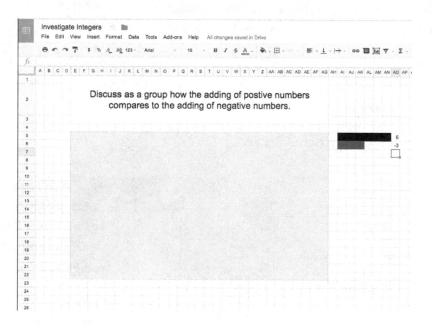

Google Classroom

Attach the small group investigation assignment to Google Classroom. Attaching as "Make a copy for each student" will result in extra documents being created since only one per group is needed. To limit confusion about which copy to use, have one student in each group share their copy of the template with group members. All group member names be added to that document title. Then, have everyone else rename their copy of the template to "Blank" to indicate that those documents were not used.

Alternatively, you can use Alice's Group Docs script which allows groups of students to be automatically added to a single document. In the spreadsheet template, paste the names and email addresses of students. If desired, use a code in Column C for pre-made groups. Use the Add-on menu to choose "Group Docs Maker." The sidebar provides a choice to create random groups or pre-determined groups. Pressing the button in the sidebar prompts you to choose a template document from Google Drive. The document will be copied and group members added as collaborators. Other students do not have access to group documents they are not a member of.

Attach the spreadsheet with the list of student groups in Google Classroom as "Students can view file." Using Group Docs Maker, there is no need to attach the group templates into the Google Classroom assignment. Each group's individual document is easily accessible to the teacher and students from the Group Docs Maker spreadsheet.

Explore More

Group Document Template:
alicekeeler.com/groupdocs

	A	B	C	D	E	F	G
1	Group Size	5	Number of Groups	10			
2	Use the Add-on menu to choose "Group Maker"						
3	Roster	Student Email Address	random	Group			
4	Lang Smedley	ss22@fresnoteach.org	0.8187342313	1	https://docs.google.com/spreadsheets/d/1iP8xPYVi		
5	Katheleen Cousins	ss49@fresnoteach.org	0.06220393737	1	https://docs.google.com/spreadsheets/d/1iP8xPYVi		
6	Sammy Halas	ss26@fresnoteach.org	0.1455143131	1	https://docs.google.com/spreadsheets/d/1iP8xPYVi		
7	Tosha Kroeker	ss44@fresnoteach.org	0.9811437633	1	https://docs.google.com/spreadsheets/d/1iP8xPYVi		
8	Peggy Dahmen	ss2@fresnoteach.org	0.5090205063	1	https://docs.google.com/spreadsheets/d/1iP8xPYVi		
9	Karla Miller	ss20@fresnoteach.org	0.592741825	2	https://docs.google.com/spreadsheets/d/1NKdqB8c		
10	Lula Enochs	ss23@fresnoteach.org	0.6745350064	2	https://docs.google.com/spreadsheets/d/1NKdqB8c		
11	Normand Marquez	ss43@fresnoteach.org	0.2822943301	2	https://docs.google.com/spreadsheets/d/1NKdqB8c		
12	Kenyatta Phelan	ss11@fresnoteach.org	0.7160671963	2	https://docs.google.com/spreadsheets/d/1NKdqB8c		
13	Danette Shoemake	ss42@fresnoteach.org	0.2957740432	2	https://docs.google.com/spreadsheets/d/1NKdqB8c		
14	Vicki Madkins	ss12@fresnoteach.org	0.6534176708	3	https://docs.google.com/spreadsheets/d/1k1K6079F		
15	Thu Towns	ss18@fresnoteach.org	0.8184912963	3	https://docs.google.com/spreadsheets/d/1k1K6079F		
16	Babette Clevenger	ss5@fresnoteach.org	0.8703249276	3	https://docs.google.com/spreadsheets/d/1k1K6079F		

21. Discuss Strategies

It is important for students to approach a problem with *strategies* rather than procedural steps. Strategies help them make connections when confronted with new situations. So, instead of asking for the solution, ask students to explain the strategy they used to find the solution. When a variety of strategies are possible, encourage students to compare and discuss their strategies with peers.

Google Slides

In the grouping dots activity below, students model their grouping strategy by dragging the dots. Students choose how they will represent their strategy. Multiple approaches are possible which opens the door for some great conversations.

The Slides activity then asks students to create their own dot groupings. A pile of dots are available "off canvas" to allow students to drag the dots to create their own activity slide. Sharing the slides with a peer, the student gets to see what strategies the peer uses.

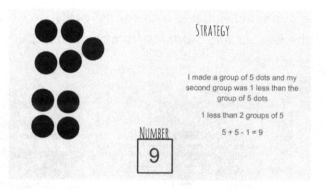

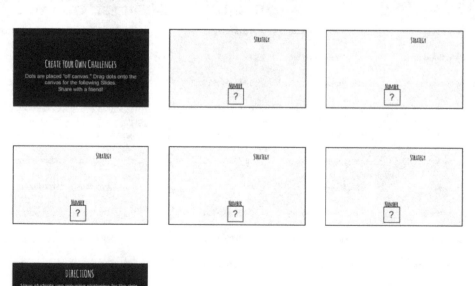

Increasing the DOK levels in math problems involves students discussing their thinking and strategies. Using Google Slides, students can include visuals to communicate their different strategies. In the probabilities with spinners activity below, students are asked to select three spinners that meet a certain criteria. The students can then duplicate the spinners of choice, remove the ones they did not choose, and use the pieces of the spinners to model their justification.

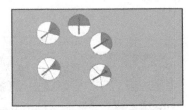

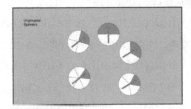

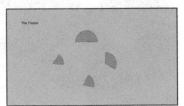

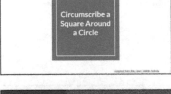
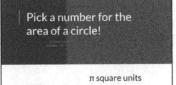
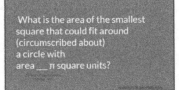

Diana's Corner

Often, an activity like Dots becomes a great intervention activity for students who have difficulty with a concept. Look for engaging activities to help create foundational knowledge.

Explore More

Grouping Dots Template:
alicekeeler.com/donotcountdots

Probability with Spinners (adapted from Open Middle activity):
goo.gl/krTeWl

22. Explain 3 Ways

To help students focus on the concept rather than just the procedures, ask students to show a single problem solved multiple ways. It is not uncommon for students to think there is one way to solve a math problem. Using this practice on a regular basis helps students learn to approach mathematics with an open mind toward the various strategies they and their classmates may use to solve problems.

Denis Sheeran's Corner

When a teacher asked her students to show how they solved a math problem in three different ways, it reminded me of cooking shows I watch where food is served "Three Ways." The variety of solutions your students bring can be surprising, interesting, and eye-opening. Give them the chance!

Problem Solving 3 Ways

Two Cookies Left

The Question

There was a jar of cookies on the table. Becky was hungry because she hadn't had breakfast, so she ate half the cookies. Then Jill came along and noticed the cookies. She thought they looked good and, so she ate a third of what was left in the jar. Denise came by and decided to take a fourth of the remaining cookies with her to her next class. Then Jodi came dashing up and took a cookie to munch on. When Megan looked at the cookie jar, she saw that there were two cookies left. "How many cookies were there in the jar to begin with?" she asked.

First Way To Solve The Problem

Working Backward:

First Way Reflection

Second Way To Solve The Problem

Second Way Reflection

Third Way To Solve the Problem

Third Way Reflection

Denis Sheeran is a Google Certified Educator and author of *Instant Relevance, Using Today's Experiences to Teach Tomorrow's Lessons.*

Templates, ideas, lessons and activities can be found at his website, denissheeran.com

Contact Denis with questions or ideas on twitter @MathDenisNJ or email denis@denissheeran.com

Google Slides

In showing one problem three ways, students can present the problem on the first slide. Each separate approach is done on a new slide. This helps clearly communicate the three different approaches.

Explore More

Denis Sheeran's Show Three: Ways Template: **goo.gl/o9J52E**

Denis Sheeran's Problem Solving: 3 Ways Example: **goo.gl/Y3zPqG**

More Problem Solving 3 Ways Samples: **goo.gl/sgzvkh**

Show 3 Ways

* Required

Get to 10

Describe 3 different ways to get to 10.

Method 1 *

Your answer

Method 2 *

Your answer

Method 3 *

Your answer

SUBMIT

Never submit passwords through Google Forms.

Google Forms

Using Google Forms, students can explain their thought process in a paragraph format. Google Forms enables students to focus more on the strategy rather than the steps to the solution since they must answer with a written response.

Google Forms makes it easy to see multiple student responses at once when you view the answers in the connected spreadsheet. You will be able to see the different strategies students used. You can also hide the column in the spreadsheet that contains student names and share the collective responses with the class.

Alice's Corner

Google Forms is assessment bliss. Seeing all student responses in a single spreadsheet makes it not only is easier than individual documents, but allows me to see patterns.

	A	B	C	D
1	Timestamp	Method 1	Method 2	Method 3
2	9/17/2016 16:06:34	5+5	1x10	30-20
3	9/17/2016 16:07:15	3 + 3 + 3 + 1	50 / 5	2 ^ 3 + 2
4	9/17/2016 16:08:03	3^2+1	15-5	20/2
5	9/17/2016 16:09:33	I can use adding to get to 10. 3 + 3 + 4 = 10	If I had 10 marbles I could group the marbles into 2 groups of 5. 2 x 5 = 10	If I had a dozen eggs and dropped 2 on the floor, I would only have 10 eggs. 12 - 2 = 10

Individual feedback can be provided to students by using the Google Sheets Add-on Flubaroo.

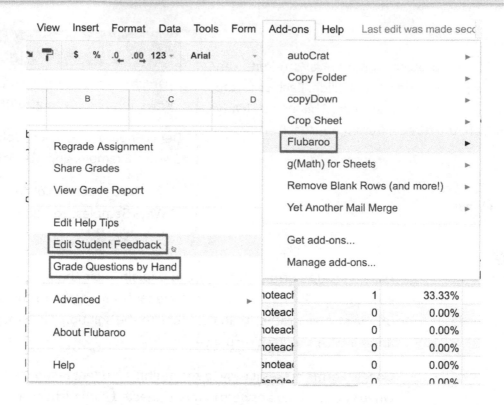

Flubaroo - Grade Questions by Hand

1. Select Student: `<<` `>>`

First Name	Jen
Last Name	F
Email Address	ss2@fresnoteach.org

2. Select Question:

Method 1 ⇕

3. Read Student's Submission: (review answer key)

5+5

4. Enter Notes for Student (sent in email):

This is one way to get to 10, can you describe your strategy along with your math problem please.

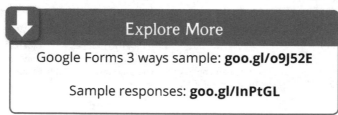

Explore More

Google Forms 3 ways sample: **goo.gl/o9J52E**

Sample responses: **goo.gl/InPtGL**

Google Sheets

Use the tabs in Google Sheets to allow students to explore and present three different models for the same problem.

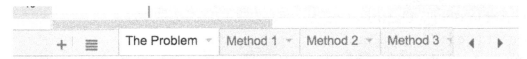

23. Create Collaborative Google Slides

Creating collaborative Google Slides is easy yet transformative. Students all type on the same Google Slides presentation. Each student adds their own slide. This makes it easy for students to "critique the reasoning of others." Students jump to another slide in the presentation and insert comments. Assessing student work is easier when it is one document rather than 30 individual documents.

1. Create **Google Slides** presentation document.

2. Name the document.

3. Close the document.

4. Share the collaborative Slides presentation with students, providing edit access.

Students access the Google Slides presentation and click on the plus icon in the toolbar. Creating blank Google Slides allows students to choose the slide layout and how to best communicate their ideas.

> **Explore More**
>
> Create Collaborative Google Slides in 40 seconds: **goo.gl/kiuTAZ**
>
> Collaborative Google Slides Template: **alicekeeler.com/slides**
>
> Collaborative Group Slides Template: **alicekeeler.com/groupslides**

Google Sheets

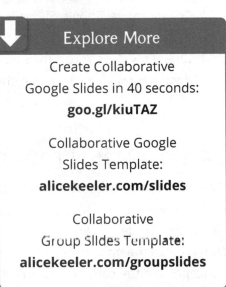

Google Sheets spreadsheets allows for each student to have their own sheet within a single spreadsheet. One document makes it easier for the teacher to give feedback as well as for students to provide feedback. Students are able to add their own tab to the collaborative spreadsheet by pressing the plus icon in the bottom left.

> **Explore More**
>
> Use templateTab to create a tab per student along with a copy of a graphic organizer: **goo.gl/N8Fec5**

Google Classroom

After creating the blank Google Slides or Google Sheets, attach it to an assignment in Google Classroom as "Students can edit file." This quickly distributes the collaborative document to the class.

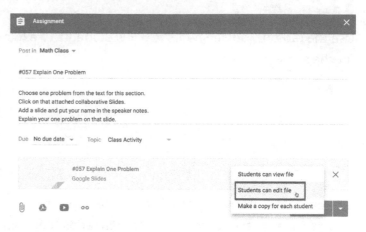

Alice's Corner

Collaborative Google Slides is one of my favorite activities to do with students. They are easy to make and truly transform the culture of my class. As often as possible, I try to use one Google Slides document with each class rather than individual documents.

24. Do a Math Problem Wrong

One way to increase critical thinking for students is to ask them to review math problems that were done *wrong*. Discuss the problem and have students critique—not criticize—their peers' reasoning. Students can view and share ideas to strengthen one another's work. This discussion creates the opportunity for students to grow in their understanding of the mathematical concepts and the communication of the concepts.

Diana's Corner

Showing problems that have been incorrectly solved is a great way to address common errors. Having students address that something is incorrect and discuss it helps them understand what went wrong and helps them remember how to find the correct answer in the future.

Google Slides

Have students work out math problems on quarter sheets of paper. Then, ask them to share a problem they did incorrectly by adding a blank slide to a collaborative Google Slides presentation. Using the Insert menu to insert an image, students can add a picture of the incorrectly solved math problem. Peers comment on each other's reasoning by inserting comments on the slide.

This is a great opportunity to help the students to understand the difference between procedural feedback and conceptual feedback. The goal is to get students to offer feedback that demonstrates an understanding of the *concepts*. Focusing on concepts rather than simply procedures may be challenging for students, so reply to student comments and encourage them to pose their critiques from a conceptual point of view.

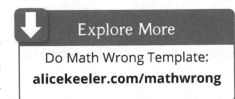

Explore More

Do Math Wrong Template:
alicekeeler.com/mathwrong

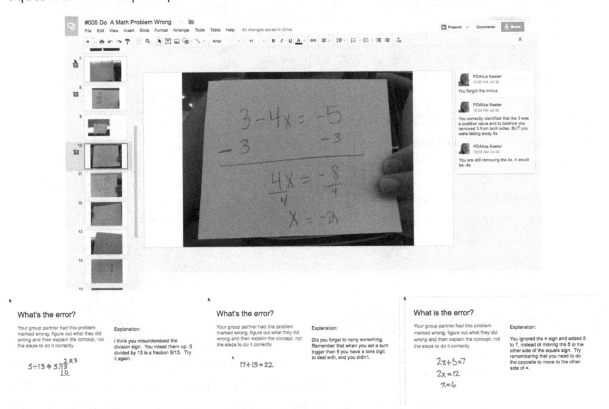

What's the error?

Your group partner had this problem marked wrong, figure out what they did wrong and then explain the concept, not the steps to do it correctly

Explanation:
I think you misunderstood the division sign. You mixed them up. 5 divided by 13 is a fraction 5/13. Try it again.

$5 \div 13 \rightarrow 5\overline{)13}$

What's the error?

Your group partner had this problem marked wrong and then explain the concept, not the steps to do it correctly

Explanation:
Did you forget to carry something. Remember that when you set a sum bigger than 9 you have a tens digit to deal with, and you didn't.

$17 + 15 = 22$

What is the error?

Your group partner had this problem marked wrong, figure out what they did wrong and then explain the concept, not the steps to do it correctly

Explanation:
You ignored the = sign and added 5 to 7, instead of moving the 5 to the other side of the equals sign. Try remembering that you need to do the opposite to move to the other side of =.

$2x + 5 = 7$
$2x = 12$
$x = 6$

25. Create a Formative Assessment Quiz

If the computer can grade it, it should.

A teacher's value lies not in the ability to grade basic skills, but rather in the ability to encourage students, provide high-quality feedback, and increase critical thinking. For DOK 1 and DOK 2 tasks, consider using applications that do the grading for you and can even provide students immediate feedback.

Google Forms

If you use a Google Form as a daily sign-in sheet, you can add warm up questions and provide immediate feedback on responses. Here's how: Include a question on the daily sign-in Form. Questions can be changed daily on the form, or you can simply leave a blank space on the form and post the questions on a Google Doc or Google Slides presentation.

Google Forms is a survey tool; however, quiz grading functions can be turned on within the Form. To do that, click on the Settings cog in the upper right corner of the Form Edit screen and select the "Quizzes" option. Turning on Quizzes allows teachers to create an answer key which creates a row in the responses spreadsheet to compare each student's answer against.

While formative assessment should not go in the gradebook, this feature can be helpful for students to do a quick check for understanding.

Explore More

DOK and learning objectives quiz:
goo.gl/NL6PO9

Branching Google Forms quiz:
goo.gl/2EtwG6

Google Forms skills quiz:
goo.gl/3bL95X

Sample Form with Flubaroo:
goo.gl/sZ5j3y

Template to create branching Google Forms:
alicekeeler.com/branchingform

When using multiple choice questions, a branching style of quiz can be created. Create multiple sections within the Google Form and choose to "Go to section based on answer." This allows for the student to get corrective feedback and a chance to answer another question before going to the next question.

There are many tools other than Google Forms that will grade DOK 1 and DOK 2 level questions for you. It is a win-win when students receive immediate feedback and the teacher's time can be spent doing things other than grading against an answer key.

Share to Classroom

Many third-party products that provide students immediate feedback work with Google Classroom. Look for a share option to "Share to Classroom."

- Kahoot: getkahoot.com
- Quizizz: quizizz.com
- IXL: ixl.com
- OpenEd: opened.com
- CK12 Practice Problems: goo.gl/TjAP0B

Non-Google Products

- Quia: Quia.com/web
- Quizlet: quizlet.com
- Nearpod: nearpod.com
- That Quiz: thatquiz.org
- Formative: goformative.com

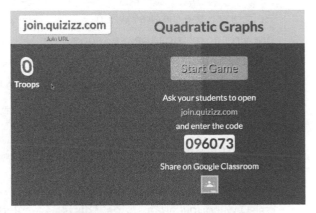

Explore More

Quizziz Help Page for Sharing with Google Classroom:
goo.gl/6nTDJq

26. Create Geometry Constructions

Creating constructions, such as the perpendicular bisector, helps students understand the properties of triangles and quadrilaterals. Constructions link to an understanding of polygon properties. Google Apps are not the only online tools available for creating constructions (euclidthegame.com is another great tool), however, using Google Apps allows for students to explain and discuss their work as well as to receive comments on their work.

Diana's Corner

Teaching constructions is a great activity for students to learn about the properties of triangles, quadrilaterals and circles. I ask students what properties ensure the construction is correct and rigid. This is when my students really start to understand all the triangle and quadrilateral properties.

Google Slides

Use Google Slides to have students create constructions. Instead of a compass and ruler, students can use the drawing tools within Slides. Holding down the SHIFT key while drawing a line emulates a straight edge. The pie-shaped icon creates a partial circle that identifies the center of the circle which can replace the compass.

Tricks for creating constructions:

- Hold down the SHIFT key when drawing or resizing the pie shape.
- Use the Tab key to cycle through elements on the slide.
- Use the arrow keys to move shapes.
- Hold down the SHIFT key while using the arrows to nudge one pixel at a time.
- CTRL-D will duplicate a shape. This is particularly handy when arcs of the same radius are needed.

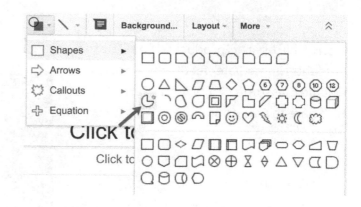

Duplicating slides allows students to show each step in their construction process. Inserting text boxes onto the slides allows students to explain each step rather than simply complete each step.

Explore More

Constructions activity in Google Slides:
alicekeeler.com/slidesconstructions

Tip To duplicate a slide, select the slide in the film-strip on the left-hand side of the screen and hold down CTRL-D.

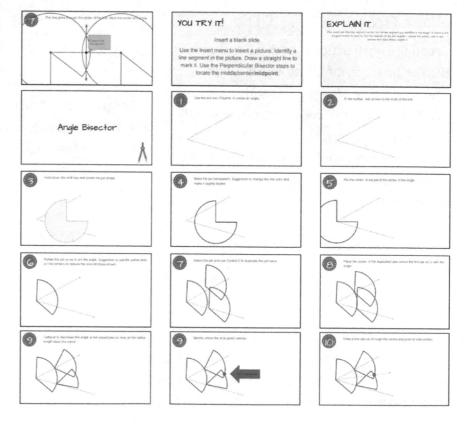

Google Drawing

Google Drawing has the same drawing capabilities as Google Slides and is a good choice when only one construction is needed.

Google Drawings easily publish to the web. The published link is an image link which allows for inserting into a spreadsheet with the formula =image("URL") where the URL is the published link from the Google Drawing.

Google Docs

Creating constructions in Google Docs allows the students to explain their thought process along with the construction. The Insert menu allows students to insert a "Drawing." The construction can be created within Google Drawings, and then the drawing can be inserted and copied and pasted within the Google Doc to allow students to build upon their constructions.

Explore More

Sample Geometry Construction in Google Docs:
goo.gl/u1T8GW

Geogebra

Geogebra math apps (goo.gl/lr6zNb) allow students to create Geometry constructions digitally. Students can submit their constructions to Google Classroom using a provided share link.

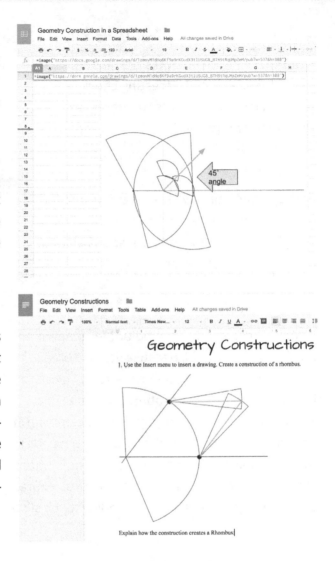

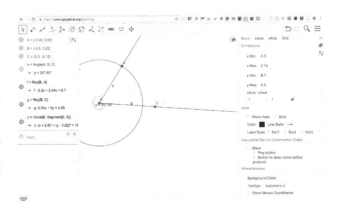

27. Create Interactive Instruction

Simply copying down notes or steps limits the amount of critical thinking a student does during instruction. A more interesting choice is to structure a Google Doc or Google Slides presentation to allow for students to interact with the information. Include opportunities for students to search for information as well as opportunities for collaboration.

Google Docs

Rather than simply reading information, students can interact with information in a Google Doc by answering questions embedded within the document.

Google Slides

Google Slides is a good medium for student math work in general, because you can break down each task the student is to do in an interactive format. Collaboration in Slides works well since all group members can contribute to the same slide at the same time. Using Daum equation editor (goo.gl/vIR2b) students can take screenshots of mathematical expressions when this is essential for adding to the slide.

Explore More

Sample Interactive Directions Document:
goo.gl/Jx6Rxm

Tips for Creating a Math Google Doc:
goo.gl/FxShdk

Skittles Lab Activity (adapted from activities by Chelsea McClellan and Fawn Nguyen): **goo.gl/7Kvava**

Explore More

Probability and Percentages Activity:
goo.gl/uc0VLB

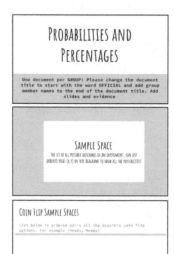

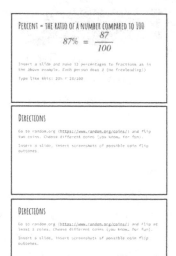

Google Classroom

The interactive document attached in Google Classroom incorporates the directions for the task. Instead of attaching in Google Classroom with view access, select "Make a copy for each student." From the Stream, students click on "Open" to view their copy of the interactive instruction document. Students are able to respond to the instruction directly on the document.

If Google Classroom is not available, share Google Docs with students through your Learning Management System (LMS) or class website. Changing the /edit at the end of a URL to /copy prompts the student to make a copy of the document rather than view the document.

Pear Deck™

One way to create interactive instruction is the use of Pear Deck (peardeck.com). Pear Deck's format allows you to create questions using draggable™ pins, drawing, text, number, and multiple choice questions. For the draggable item, create an image that students can move, such as a point or figures for a Venn diagram. You can also create a canvas for students to draw on, such as a coordinate system or graphic organizers.

28. Create a Drawing

Mathematics is visual. When students create drawings and models of their mathematics in their work the concepts become tangible to them.

Google Drawing

Using a Google Drawing document, students can collaboratively create a flowchart, infographic, meme, manipulative activity, or other visual representation of the mathematics concept they are learning.

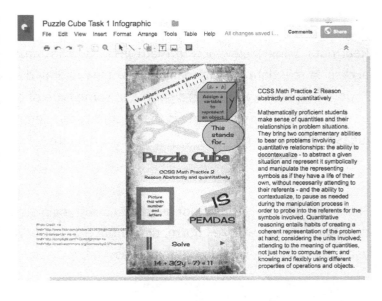

Tips for creating Drawing elements:

- Hold down the SHIFT key when drawing a circle or other shape to constrain the proportions (make a perfect circle.)
- Hold down the SHIFT key to draw straight lines.

- Hold down the SHIFT key when rotating shapes to rotate in 15-degree increments.
- Hold down the V and click on elements in the drawing to select them together.
- Draw shapes "off canvas" to create elements that students can bring into the document.
- Use CTRL-D to duplicate shapes.
- Hold down the Shift key when using the arrow keys to nudge shapes one pixel at a time.
- Use the Arrange menu to align shapes.
- Formatting one shape and drawing another shape while the first one is selected will match the formatting on the second shape.

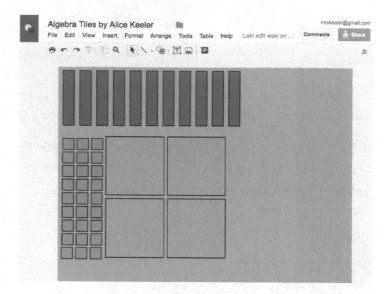

Explore More

Google Drawing flowchart:
goo.gl/pi5pZ1

Meme template in Google Drawing:
alicekeeler.com/meme

Interactive Algebra Tiles:
alicekeeler.com/algebratiles

29. Use Real Data

Real data brings relevance to math lessons. With easy access to the Internet comes easy access to real data. Traditionally, students would all do the same problems out of the book. Today, almost any website is a source of some type of data. A benefit to using the Internet as a resource for real-world data is that students can demonstrate creative thinking by exploring ideas that help them personally connect with math.

Google Flights

Flights.google.com helps students to discover new places. Students can each choose different destinations and explore information about that place. The information from the flights can be used to demonstrate learning objectives.

Google Search

Making math relevant can be as simple as using real product examples. Using the price and other data from actual products that are currently available for sale, students can make calculations around real products. With a quick Google Search, students can find product pricing and create and answer their own math problems. No more contrived word problems!

Explore More

Online Shopping Sample
by Denis Sheeran:
goo.gl/Zx2ngj

Upgrade the "Coffee
Math" Worksheet:
goo.gl/iTdUq8

Design a Menu by
John Stevens:
goo.gl/9ztHPR

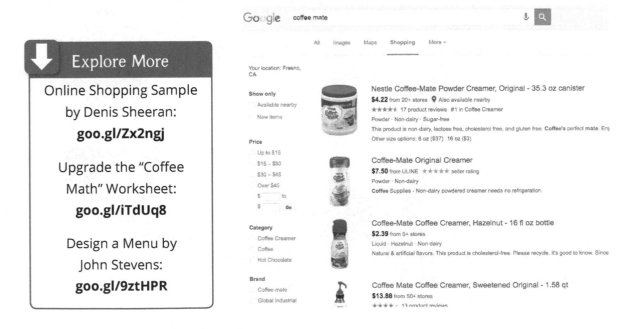

30. Use g(math) in a Google Form

The g(math) Add-on for Google Forms allows you to create graphs and make mathematical notations within a Google Form. To install this Add-on, click on the three dots menu in the upper right of the Edit screen of Google Forms and choose "Add-ons."

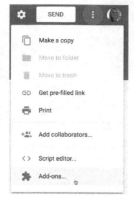

Locate g(math) and add the Add-on. The next time you create a Form and use g(math), look for the puzzle piece icon in the upper right and choose "g(Math) for Forms."

Choose from a selection of mathematical question types. "Create math expressions" allows you to insert images of mathematical expressions by using voice commands or typing with calculator math.

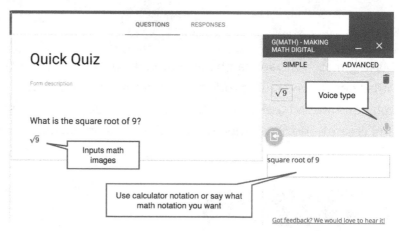

In addition to mathematical expressions, g(Math) will insert graphs into the Google Form.

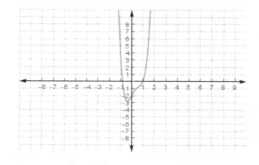

Answer for

Your answer

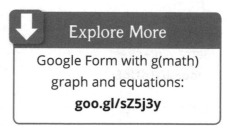

Explore More

Google Form with g(math) graph and equations:

goo.gl/sZ5j3y

31. Create Video Playlists

Students are no longer restricted to learning at a set time and location. Videos can be a great way to transform, or even flip, the math classroom. Rather than creating one long video, create a set of short videos to show each step of the math process. Try to keep videos to about thirty seconds each. For students, short videos are easier to view and *re*-view until they understand the concept. For teachers, it's easier to maintain enthusiasm for the length of a short video. Shorter clips are easier to edit and update.

 Tip: Remember, the students can replay the video. You don't need to repeat information, pause, or talk slowly.

YouTube Live

YouTube allows anyone to create events, record them, and save them directly on the platform. Go to youtube.com/my_live_events to schedule an event. This page can also be accessed by clicking on the user avatar in the upper-right corner and choosing "Creator studio." Events are located under "Live streaming" on the left-hand side.

Students can use YouTube Live to host collaborative podcasts to explain math concepts. Teachers can use this feature to host office hours or a study group event or to record a review of math concepts. Since the videos are saved in YouTube, they can easily be gathered into a YouTube playlist for sharing.

YouTube

Creating playlists of videos in YouTube allows for a mix of teacher-created and student-created videos. You can even select and add math videos from others that are available on YouTube. Playlists can also be created from a single user's video library.

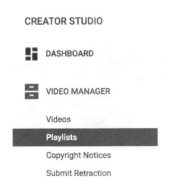

When viewing a video on YouTube, click the "Add to" option beneath the video. This allows you to add a video to an existing playlist or create a new playlist.

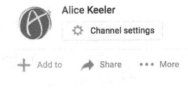

Playlists are accessible from the "Creator studio" at youtube.com/view_all_playlists. Videos in a playlist can be reordered and shared as a set of videos.

Google Drive

Videos in Google Drive for G Suite accounts do not count against your Google Drive storage. Creating a folder in Google Drive with videos is a way to create a video playlist without YouTube restrictions or ads. Share the folder by clicking on the link sharing icon in the Google Drive toolbar. Copy this link and distribute.

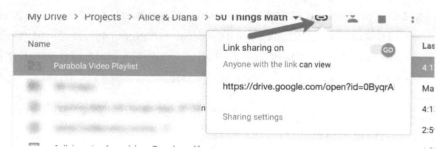

When creating a list of videos in Google Drive, number the videos so they can be easily sorted. Sorting the folder by title will order the videos accordingly. Using a two-digit numbering allows for proper alphabetizing of the video titles. Titling the videos with very specific information as to what the video contains helps students to more efficiently choose the video they need.

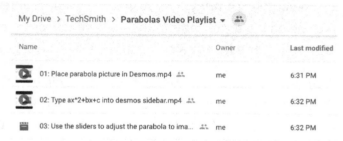

Google Classroom

YouTube playlists or Google Drive playlists can be easily shared in Google Classroom. Paste the link to the YouTube playlist or Google Drive folder into a Google Classroom assignment so students can quickly connect to the playlist.

Google Classroom makes it easy to create additional Google Classroom classes that can be used as a resource repository. Playlists created in a Google Classroom assignment can be a mix of videos from YouTube and Google Drive.

1. Create a class just to house video playlists.

2. Attach the videos in the intended order of viewing.

3. Invite the students to the video repository class.

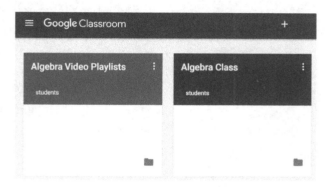

This allows students to locate playlists of videos on topics they need help with.

Each "assignment" in the video resources class can be easily linked to. The three dots icon in the upper right of each assignment offers the option to "Copy link." Assignments in the primary Google Classroom class can link to the video playlist as one of the attached resources. When creating an assignment, the link icon allows the playlist link to be pasted.

32. Show Me You Know It

Traditional math assignments that ask students to complete a math problem and show the steps are less relevant in an age of Google and mobile apps, where the answers and the processes can found with a quick search. Even when students work out these problems without the aid of technology, oftentimes the problems are DOK 1. Students may be able to follow steps, but they may lack the ability to explain their thinking process or recognize how to apply it.

Students need to demonstrate a deep understanding of math concepts. Rather than avoiding the use of apps that provide students with steps, change the style of questioning so that students are encouraged to think more critically. For example, have them approach a problem from multiple directions as a way of showing understanding and not simply computational ability. Students can personalize their understanding by identifying a situation where the concept can be used in activities they are involved in or in current events. Asking students to "show me you know it" asks the students to increase their level of depth of knowledge from DOK 1 to DOK 2 or DOK 3.

Activity	Example
Create a mathematical "Would you rather" activity.	These activities don't necessarily have one correct answer; instead, it is all about their justification of their answer. For example: Would you rather be a triangle or a square? Would you rather be a solution to a quadratic or the solution to a linear equation?
Compare procedures used in solving problems.	Instead of just assigning isolated problems, ask students to compare the solving processes for two or three of the traditional problems. Discuss what is similar with the processes and what is different to make them unique.

Activity	Example
Use Open-Middle (openmiddle.com) for a group activity.	Students in the group organize into recorders and thinkers. Thinkers talk through the process, and recorders write down the thoughts.
Create a campaign speech for the topic.	You are a triangle (or any other shape). Create a campaign speech as to why you should be voted the best polygon. You are the digit 0. Create a campaign speech on why you should be the most important digit.

Google Slides

Students can work in groups using a collaborative Google Slides presentation to organize and present their ideas and defend their arguments. Here are a few tips for using Google Slides:

- Have one student create a Google Slides presentation and then share it with the group's other members.
- In Slides, students can use the shapes tool to model their mathematics.
- Side–by-side comparisons can be created by opening the drawing tool in Slides.

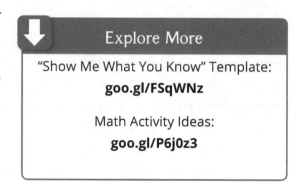

Explore More

"Show Me What You Know" Template:
goo.gl/FSqWNz

Math Activity Ideas:
goo.gl/P6j0z3

33. Use Spreadsheet Formulas

Using Google Sheets allows students to conceptualize their math, write formulas to demonstrate understanding of the steps, and use formulas to check their work. When the learning objective is not to calculate by hand, spreadsheets can facilitate calculations to allow students more time to think and analyze.

Spreadsheet formulas emphasize process over solution. The teacher is able to use the View menu to choose "All formulas" to reveal that students wrote formulas to calculate their values. This provides insight into a student's understanding of concepts and algorithms.

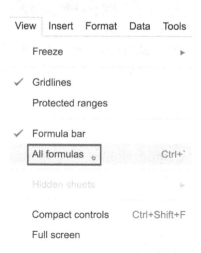

Google Sheets

Ten frames are a way for students to conceptualize the value of ten. For this example, transitional kindergarten (TK) teacher Christine Pinto used Google Sheets to have four year olds demonstrate their understanding of the value of ten.

Students are provided a graphic organizer in the spreadsheet to explore the concept of ten. Conditional formatting in the spreadsheet allows the students to type a single digit number that corresponds to a color. This allows the students to fill in their ten frame on the spreadsheet. The constraint is that students can only use two colors to fill the ten frame. Students count each color used in the ten frame and type an equation equaling ten in an answer cell.

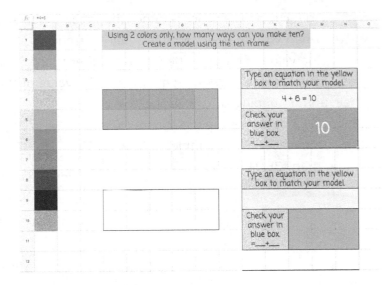

To check their work, the TK students type a formula. To type a formula into a spreadsheet, students start by typing an equals sign (=) followed by summing the two numbers they counted. When they press enter or an arrow key, the spreadsheet will validate that their sum indeed adds to ten. The next tab in the spreadsheet challenges the student to go further by using the constraint of three colors.

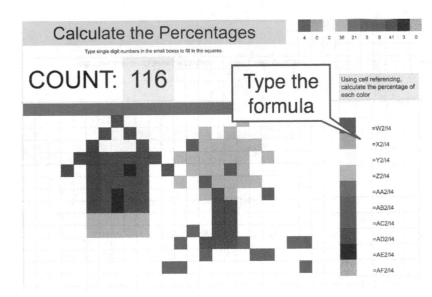

In teaching students concepts such as percentages, the learning objective is not to demonstrate the standard algorithm for division. Students can type formulas into the spreadsheet to demonstrate that they know how to perform the calculation for percentages. Students would then explain and interpret each percentage.

To focus on the process rather than the numbers, students should use cell referencing when writing their formulas. Cell referencing in a spreadsheet is when students use the value of a cell in the formula rather than the numerical values. If the spreadsheet has a count of the parts of something and a value of the whole, students can demonstrate their understanding of calculating the percentage by typing the equation into the spreadsheet. For example, =W2/I4 rather than =36/116.

Diana's Corner

For secondary levels, my students enjoy creating "targets" for each other to determine the probability of hitting one of the colors.

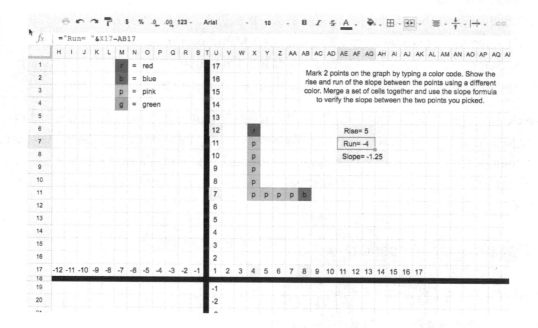

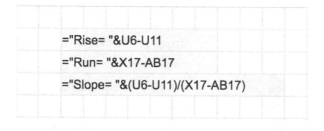

Explore More

10 frame activity by Christine Pinto:
pintomake10.gafe4littles.com

Percentages pixel activity:
alicekeeler.com/pixelpercent

Slope Activity: **goo.gl/8Rgyk5**

34. Geoboard Activity

Geoboard is a virtual manipulative available in the Chrome Web Store (goo.gl/PhTdYC). Students can use this manipulative to deconstruct a shape into simpler shapes in order to calculate the area. Students create a challenging shape in Geoboard and then calculate the area two different ways. After calculating the area of the shape, students share their shapes with other students.

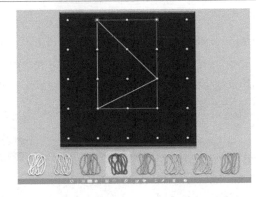

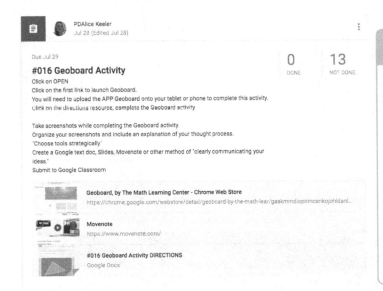

PDAlice Keeler
Jul 28 (Edited Jul 28)

Due Jul 29

#016 Geoboard Activity

	0	13
	DONE	NOT DONE

Click on OPEN
Click on the first link to launch Geoboard.
You will need to upload the APP Geoboard onto your tablet or phone to complete this activity.
Click on the directions resource, complete the Geoboard activity

Take screenshots while completing the Geoboard activity.
Organize your screenshots and include an explanation of your thought process.
'Choose tools strategically'
Create a Google text doc, Slides, Movenote or other method of "clearly communicating your ideas."
Submit to Google Classroom

Geoboard, by The Math Learning Center - Chrome Web Store
https://chrome.google.com/webstore/detail/geoboard-by-the-math-lear/gaakmrndiopnmcenkojohldani...

Movenote
https://www.movenote.com/

#016 Geoboard Activity DIRECTIONS
Google Docs

Diana's Corner

The Geoboard app has long been my students' favorite app. They create shapes and challenge one another to determine the area. From my viewpoint: no rubber bands flying through the classroom :)

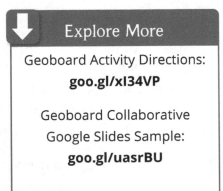

Google Slides

As with other activities, students can use Google Slides to share their manipulatives and comment on the activities of other students. Screenshots taken during the process, as well as of the final work, can be dragged or inserted into the Slide.

Explore More

Geoboard Activity Directions:
goo.gl/xl34VP

Geoboard Collaborative Google Slides Sample:
goo.gl/uasrBU

35. Use Manipulatives

While physical manipulatives are great, they do have some drawbacks. Access to manipulatives, storage, set up, transition time, capturing learning, and losing pieces can all be issues. Google tools allow you to create templates of manipulatives that students can use to model their mathematics. The collaborative abilities of Google tools allow students to work together using manipulatives.

Google Drawing

A Google Drawing makes the perfect canvas for manipulating objects. Like other Google products, Google Drawing is collaborative. Multiple students can work together on one drawing to model their mathematics.

Encourage students to take advantage of the area around the drawing. The "off canvas" area is perfect for placing objects for students to use in their modeling. Students can then drag items onto the canvas when they are needed for their models.

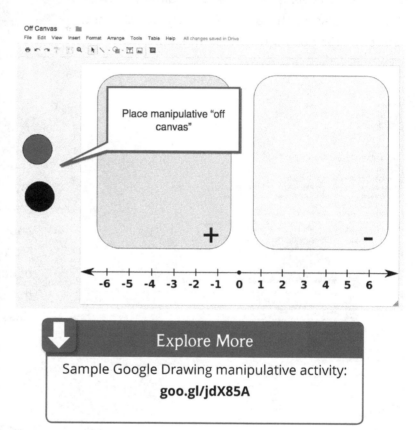

Explore More

Sample Google Drawing manipulative activity:
goo.gl/jdX85A

Google Slides

Google Slides provides the same graphical opportunities as Google Drawing. Using Google Slides allows for multiple pages of manipulatives. Unlike Google Drawing, backgrounds can be locked into place in Google Slides. Create an image for the background and click on the "Background" button in the toolbar to choose the image as the background image.

Try creating the background image on a blank slide:

1. Use the "File" menu to "Download as" and choose "PNG image (.png)."

2. Use the downloaded image as the background image on a new blank slide.

3. The image dimensions should match the slide dimensions. Avoid stretching the image.

To create a manipulative activity in Google Slides, place the manipulative elements "off canvas" from the slide. Select the slide in the filmstrip. Copy the slide and paste repeatedly to create multiple manipulative opportunities.

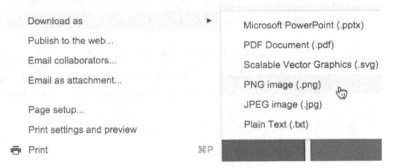

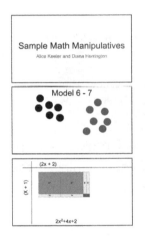
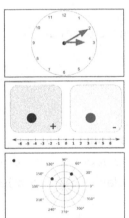
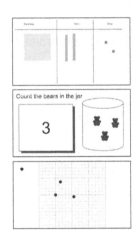

Diana's Corner

When using hands-on manipulatives, students will enjoy using their cell phones to make a video of the manipulatives.

 ## Explore More

Sample Google Drawing manipulative activity:
alicekeeler.com/ manipulatives

Coordinate Battleship Template:
alicekeeler.com/ slidesbattleship

Drag and Sort Shapes Activity:
alicekeeler.com/ dragandsort

Drag Counting Bears Activity:
alicekeeler.com/ dragthebears

Alice's Corner

Instead of using Google Slides to give information, use it to get information. I design my Google Slides with the idea that students will interact with them. Activities with dragging elements are one way to do this.

Google Docs

Manipulatives can be used within a Google Doc as well. While Google Docs are collaborative, the manipulative elements are not. This makes using manipulatives in a Google Doc more suitable for an individual report or project than for group work.

1. Use the "Insert" menu to choose "Drawing." This inserts a drawing canvas into the Google Doc.

2. Create manipulative items such as integer dots in the drawing.

Explore More

Sample Google Docs Manipulative:
goo.gl/flVKrz

3. Students can then double-click on the drawing within the Google Doc to activate the drawing canvas.

4. Students will rearrange the manipulatives to make them display properly within the Google Doc.

36. Create Collaborative Maps

Mymaps.google.com is a collaborative tool that allows students to create math projects using a Google Map. Students are able to measure on the map, identify locations, and add pins. In the pins, students can provide text or multi-media to explain the pin location. This exercise gives students real-world context for their math.

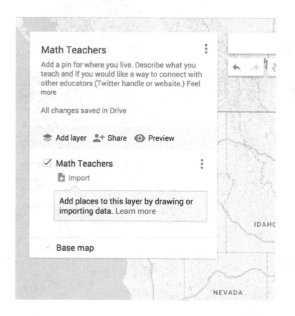

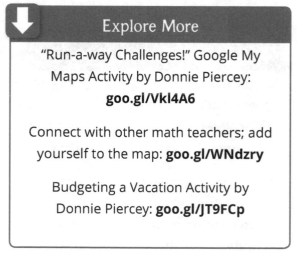

Explore More

"Run-a-way Challenges!" Google My Maps Activity by Donnie Piercey:
goo.gl/Vkl4A6

Connect with other math teachers; add yourself to the map: **goo.gl/WNdzry**

Budgeting a Vacation Activity by Donnie Piercey: **goo.gl/JT9FCp**

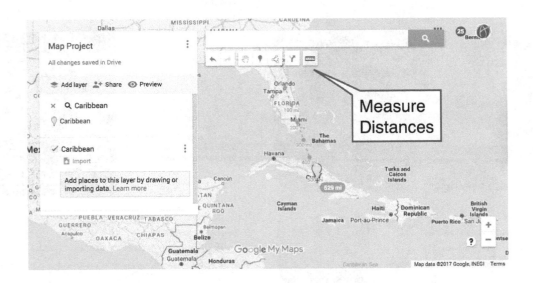

37. Students Write Guiding Questions

Teachers often give overriding questions/tasks to students, but a stronger approach is to provide a situation and have students create their own guiding questions to answer in order to complete the task. After receiving the task, have students brainstorm what they need to know to accomplish it.

During this process, students should create their own set of guiding questions. These questions ultimately may or may not be used by the students, but because they are focused questions, they will help guide students to the completion of the task.

Diana's Corner

When providing my students with a challenge I do not give them a template, I only give them a situation. They create blank slides and write their own guiding questions in order to tackle the challenge.

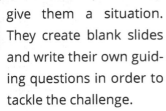

The four slides below from a final presentation show the thought process used to complete the assigned task.

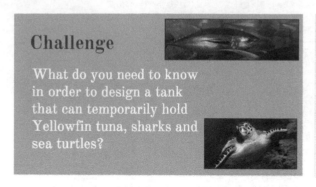

Challenge

What do you need to know in order to design a tank that can temporarily hold Yellowfin tuna, sharks and sea turtles?

Essential Questions

We asked ourselves six questions, but only used questions #3 - #6 to come up with our design.

1. How big are the animals?
2. How much space do the animals need?
3. What are the animals' turning radii?
4. What shapes are conducive to the swimming patterns of the animals?
5. What is the unit rate of gallons of water per cubic foot?
6. How deep does the water need to be for the animals?

Rationale

The Eliminated: Questions #1 and #2

- The first question we thought of was that we needed to know how big the animals were, but as we worked, we realized knowing the animals turning radii would be sufficient.
  - Size of turning radius is indicative of the size of the fish. If the fish can comfortably turn around, then the fish can also comfortably swim in a straight line.
- Initially, we designed the tank as though we were simulating the habitat of the animals and thus wanted to consider how much space the fish usually need in the wild.
  - We quickly realized we did not need to perfectly replicate the conditions the animals experience in the open ocean because the amount of space the animals actually occupy is not feasible for a tank

Rationale

The Included: Questions #3-#6

- Part of the challenge of designing the tank was to consider the size of the yellowfin tuna's turning radius.
- Not only do you have to consider how big the size of the fish's turns are, but you need to consider the amount of space needed to complete that turn.
  - i.e. Sharper corners don't provide a lot of room for fish to turn around, especially if multiple fish swim into the same corner/space at the same time.
- We knew that we had to ensure the tank could hold 1,000,000 gallons of water and thus we needed to know how much space each gallon of water would occupy so we could determine the minimum size the tank would have to be to hold a million gallons.
  - By determining how many cubic feet one gallon would occupy, we could multiply this amount by 1,000,000 to get the full size (in cubic feet) that we needed.
- We initially wanted to ensure that the animals would have enough space in the tank so we looked at the average depth that they usually dove to so we could make sure the tank we designed would accommodate this instinct. However, we quickly realized that this was not feasible and instead chose the average depth of an aquarium holding tank.

Google Classroom

Post the task in Google Classroom. Students can use the public commenting feature in the Stream to brainstorm the guiding questions. Using the Question style assignment in Google Classroom allows students to submit a question. Students can view and comment on the submissions of other students as the questions are being developed.

Google Docs

Students can collaborate on a Google Doc to brainstorm the guiding questions related to a task. Students should come back to the document (after completing the task) to reflect on which questions they used and which questions they did not use, and why. This process teaches them how to ask better questions. The sooner students learn to ask better questions, the sooner they become stronger learners.

Explore More

Guiding Questions Template:
goo.gl/8TVwwj

Google Slides

Google Slides creates an excellent platform for students to organize and display their work. The first few slides should contain the students' guiding questions, followed by their process for tackling the task. As students answer the guiding questions, they can state why they are using them. Analysis and connections are built within their problem-solving process.

Explore More

Sample Student Project using Google Slides: **goo.gl/4fkeHs**

38. Have Students Design Spreadsheets

The ability to design a spreadsheet is a valuable skill that most students will need both in and out of school. Spreadsheets can help students organize a task and analyze information with charts and graphs. Spreadsheet formulas provide ways students can demonstrate their understanding of math concepts in a meaningful context.

Google Sheets

Selected list of spreadsheet skills:

- Resize rows
- Resize columns
- Set word wrapping
- Set horizontal and vertical alignment
- Select a range of cells
- Fill in background colors
- Merge cells
- Freeze the first row
- Create column headers
- Start a formula with an equals sign
- Double click a cell to edit a formula
- Use cell referencing
- Use absolute cell referencing
- Create additional sheets to organize information
- Create a chart of data
- Utilize the Explore feature to create charts
- Analyze appropriate charts to use
- Create lookup tables
- Use fill-down patterns
- Drag and move cell values
- Paste special

Explore More

Template for Practicing Spreadsheet Skills to Organize and Display Work: **alicekeeler.com/ sheetspractice**

Using Spreadsheets to Introduce Functions Activity: **goo.gl/CbzYcb**

Spreadsheet Skills Practice: **alicekeeler.com/ advspreadsheet**

Learn Basic Spreadsheet Skills Template: **alicekeeler.com/ basicspreadsheet**

39. Analyze Data Sets

Oftentimes, students are given neat data tables rather than messy data sets. A needed skill in many modern workplaces is the ability to manage large amounts of data. You can help them develop this skill by providing the students with a real data set and asking them to analyze it for meaning. The following questions are a few that they may need to be able to answer as they work with the data:

- What needs to be cleaned up?

- Is there text in the cell that needs to be separated from the numbers?

- What units are being used?

- What is important?

- What is extraneous?

- What charts would make sense to display this data?

- What charts would not make sense and why?

- How can I (we) demonstrate the learning objective with this data?

- How can I (we) communicate the data in a way that makes sense?

Google Sheets

Research.google.com/tables is a Google Search that only returns results where the websites contain data tables. An "Export to Google Sheets" button on the search results copies the data from the website into a spreadsheet.

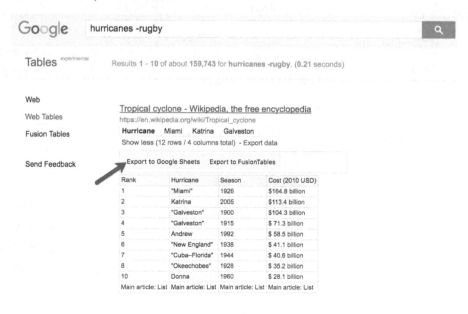

From the spreadsheet, students can highlight the data and use the "Explore" button in the bottom right to automatically create graphs. The graphs may be useful or they may be ridiculous. Students need to analyze the resulting graphs. Taking advantage of the Explore feature, automatic calculations in a spreadsheet, pivot tables, and the chart editor allows students to focus on the analysis and to think critically about the procedural steps of calculating and graphing.

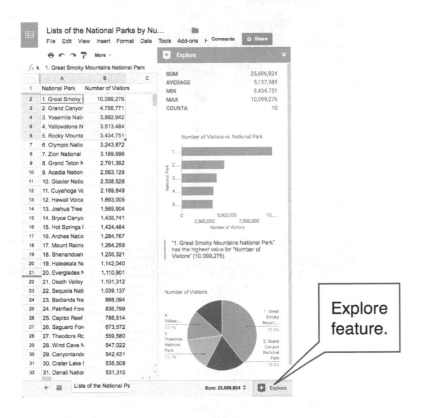

Explore feature.

Explore More

Perform Data Searches:
research.google.com/tables

Data search on the California Drought:
goo.gl/oGGbe2

Analyze Flight Costs Activity:
goo.gl/iGy6JJ

Essential Questions and Search Links:
goo.gl/BXRlG5

Alice's Corner

Almost every managerial job requires the use of spreadsheets. Having students learn spreadsheet skills gives them not only a purpose for doing math, but also a valuable life skill.

Google Slides

Students can present their data and results with Google Slides. Charts created in a Google Sheets spreadsheet can be inserted into a Google Slides document, Google Doc, or a Google Drawing. The inserted charts are linked back to the spreadsheet and can be updated when the spreadsheet data changes.

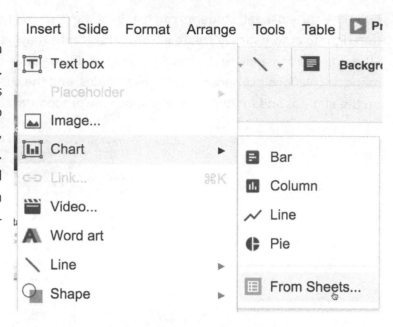

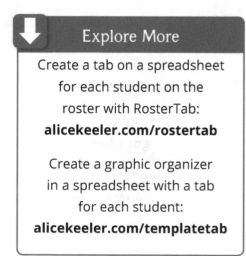

Explore More

Sample Google Slides with linked chart from a spreadsheet: **goo.gl/AyaFSf**

40. Crowdsource Information

Organizing data is a natural fit for a spreadsheet. Before collecting information for an activity, students need to create column headers and determine how they will make the best use of the sheets within a spreadsheet to group information.

Google Sheets

Since Google Sheets allow for collaboration, they are a great tool for crowdsourcing data. Students can work in small groups or as a whole class. Each student can add a sheet to the spreadsheet and rename the tab to their name. If students are collecting data from different sources, the whole class benefits from the collaborative research.

Explore More

Create a tab on a spreadsheet for each student on the roster with RosterTab:
alicekeeler.com/rostertab

Create a graphic organizer in a spreadsheet with a tab for each student:
alicekeeler.com/templatetab

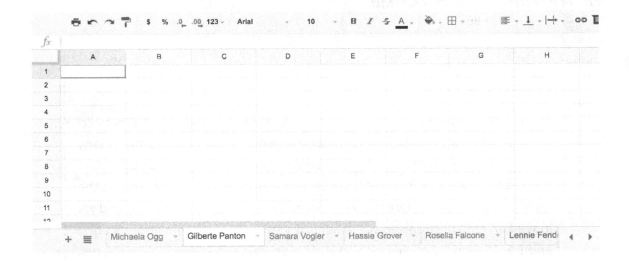

41. Focus on the Learning Objective

The ultimate goal is for students to demonstrate their learning. A demonstration of learning can come in a variety of ways. Rather than providing students pre-constructed math problems for an assessment, have them use their knowledge with a personalized adaptation. Provide students with the objectives they need to demonstrate. Then, per the mathematical practice to "choose tools strategically," encourage students to determine the tools and methods for showing their understanding of the concepts. The focus for the assessment is the main idea; you will be able to see how students understand what has been taught and how DOK levels change as they create their understanding of the lessons and show their connections.

Diana's Corner

At first the students thought this would be easy, but because it was so open-ended they took the exercise much further than I would have asked them.

Google Docs

Google Docs are an easy way to provide students with a list of elements they need to demonstrate. As per the eight mathematical practices, students can choose the appropriate tools to clearly communicate their understanding of the objectives.

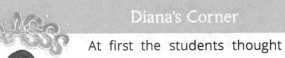

Explore More

A Sample Test with Constraints Rather than Questions:
goo.gl/89PbtE

Sample directions for an assessment:

Chapters 1–2

Assessment Task

In Chapter 1 we investigated number patterns. We looked at Fibonacci numbers, square numbers, triangular patterns, even numbers, and many more. We noticed that recurring patterns can be modeled with an equation. In Chapter 2 we looked at sets, subsets, the amount you could have in the number of subsets, different ways to identify members for a subset, and we looked at Venn diagrams as the graphic organizer for how sets intersect one another. We looked at intersection and union notation and what that means. We touched on logic notation with and/or and truth tables.

For this portion of your assessment, you are going to create your own investigation with the following parameters:

☐ Choose two of the number patterns we studied.

☐ Create a third number pattern; name it after yourself.

☐ Discuss the mathematics behind each of the three patterns.

☐ Create at least one Venn Diagram that demonstrates where the first ten numbers of each pattern exist.

☐ Discuss where the patterns overlap, if they do, and why or why not.

☐ You can include truth table information or any other additional information you would like.

In Google Classroom, click on OPEN for this assignment. Click on CREATE and create either a Google Doc or a Google Slides. Use this to complete your task.

Organize your work and communicate your ideas so it is clear to others besides the teacher. i.e.: someone who is unfamiliar with this material.

Organize your work so it is evident that you accomplished the constraints of this task.

Google Sheets

It is hard to beat a spreadsheet when it comes to creating a list. Designing rubrics in a spreadsheet helps students to ensure they address the necessary elements of the task.

Google Forms

When collecting information from students, consider using a Google Form. The convenience of viewing all student responses in a single spreadsheet along with summary data of responses makes Google Forms a very useful tool.

Use Google Forms to provide students with the list of elements assessed. Using multiple choice questions or a multiple choice grid, students can indicate if they addressed the element, are progressing on showing that element, or if they did not address the element.

In the Form settings, select the option to collect student email addresses and turn on Quizzes. For each listed element the student is asked to demonstrate, selecting "Answer key" allows for adding point values to the question. The teacher is able to review individual student responses in Google Forms, assign partial credit, and provide individual feedback per student per objective.

Explore More

Holistic Scoring Rubric:
alicekeeler.com/mathholisticrubric

Single point rubric template:
alicekeeler.com/singlepointrubric

RubricTab: assess on a four-point rubric and send a copy of the spreadsheet to each student:
alicekeeler.com/rubrictab

QUESTIONS   RESPONSES

Demonstrate Learning

In Google Classroom create a Google document (Slides, Sheets or Text Doc) to present your demonstration of the following learning objectives.

Submit Link to Digital Document

Long answer text

Choose two of the number patterns that we studied

○ Demonstrated

○ Progressing

○ Did not demonstrate

☑ Edit feedback and points:

Create a third number pattern, name it after yourself   4 points

Explore More

Number Patterns Assessment Google Form Sample: **goo.gl/K7SBID**

Number Patterns Assessment—create a copy of the Google Form: **goo.gl/UBbt2n**

Make a Google Form Template—copy and paste learning objectives into the spreadsheet to quickly make a Google Form:
alicekeeler.com/form

42. Use Graph Paper

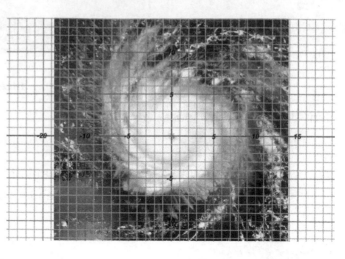

Graph paper is useful for teaching math—whether that "paper" is physical or digital. When using grids, x-y axis, etc., with Google Apps, it is helpful for images to have a transparent background (requires PNG file types). Use the images as backgrounds to build graphs on top of, or use the graph paper images on top of pictures to allow for the application of mathematical ideas.

Google Drawing

Students can use graph paper right in a Google Drawing. One way to get graphs into Google Drawing is to go to Google Images and search for "graph paper" or "x-y axis." Within the search, click on "Search tools" to filter the color for "Transparent." These images can be dragged onto a Google Drawing, with no downloading required. Pay attention to copyright and provide attribution. Remember, giving credit does not mean you have permission to use an image.

> ### Explore More
>
> Google Drive folder of graph paper drawings:
> **alicekeeler.com/graphpaper**
>
> Blank Google Drawing template sized for Google Slides:
> **alicekeeler.com/drawslides**

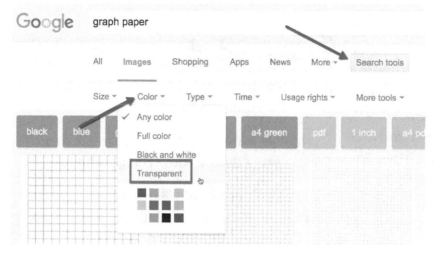

One way to make sure you have permission to use a graph image is to make your own. Graphs and grids can be created in Google Drawing. By default, Google Drawings have a transparent background. Downloading the drawings as PNG allows them to be used in Google Docs and Google Slides.

Google Slides

Modifying the slide master allows you to create graphic organizers that have a graph paper background. When students add a new slide, the slide automatically has graph paper.

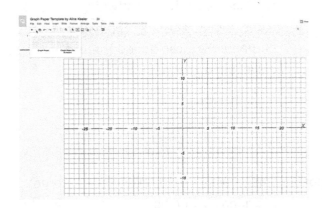

Explore More

Google Drive folder of graph paper drawings:
alicekeeler.com/graphpaper

Google Slides Template with Graph Paper Backgrounds:
alicekeeler.com/graphpaperslides

As an alternative to having the graphs locked to the background, you can add the graph onto a slide. This allows the graph to be placed over other images. If the graph paper is behind the image, use the Arrange menu to bring the graph to the front.

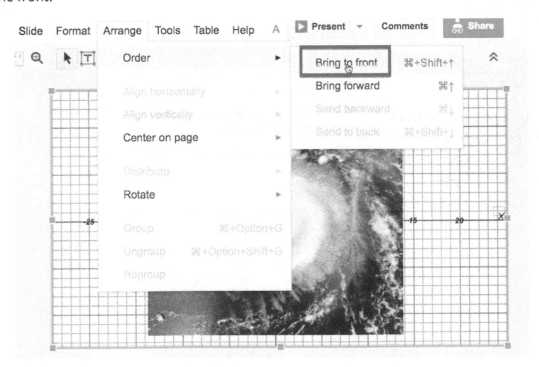

43. Collect Data

Alice's Corner

How does tech make teaching and learning better? The ability to collect data helps me to spend less time grading and more time analyzing the data so I can improve my lessons and target students for help.

In a modern age, digital data should be collected and used to help guide learning. When students are marking responses on paper or PDFs, giving feedback is time-intensive for the teacher. Receiving feedback quickly is motivating and provides more learning opportunities for students. You can speed the process by collecting digital data.

Google Forms

Using Google Forms for the multiple choice portions of an assessment allows for automatic grading and the collection of data. The item analysis provided by Google Forms can be invaluable for targeting re-teaching needs. The connected Google Sheets spreadsheet allows for data analysis for further instruction and for teacher reflection. Responses can be sorted by question to look for patterns and to quickly get a list of students who need to work with the teacher in a small group for a particular concept. Pivot tables can be generated from the student responses to provide summary data.

For questions where more than one response may be correct, the "Checkboxes" style of question permits multiple responses, whereas "Multiple choice" questions in a Google Form only permit a single response.

QUESTIONS   RESPONSES

Question 1

Description (optional)

Which of the following sequences is arithmetic with the 100th term being 5020?   ☑ Checkboxes ▼

☐ A. 1, 5, 9, 13, …   ✕

☐ B. 70, 120, 170, …   ✕

☐ C. 1, 3, 9, …   ✕

☐ D. 1060, 2000, 2040, …   ✕

☐ Did NOT choose this question.   ✕

☐ Add option or ADD "OTHER"

☑ ANSWER KEY   (2 points)   ▢   🗑   Required ◯   ⋮

Turning on Quizzes in the settings allows not only for point scoring, but also individual feedback after the student has submitted his or her response. Individual feedback can be provided on any question type except for "Multiple choice grid." In the Google Form edit screen, for Forms where "Quizzes" have been turned on, individual feedback can be left on the "Responses" tab.

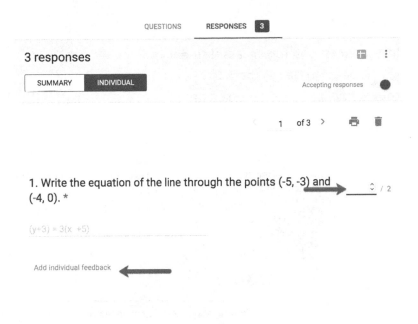

Explore More

Sample math assessment in Google Docs: **goo.gl/zFNifl**	Google Form template—Choose 5 out of 10 multiple choice: **alicekeeler.com/5of10form**
Sample Google Form multiple choice response: **goo.gl/ZaD9Ve**	Sample formative assessment by Megan Heine: **goo.gl/4pJ4VT**

44. Provide Assessment Choices

Rather than assessing what students do not know, assess what they do know. Give students an opportunity to choose from a selection of questions. This practice can be used regardless of whether the type of assessment being executed is traditional or non-traditional. For multiple choice formats, if you want students to do ten problems, have them choose from fifteen problems. On a short answer test, if you want students to answer five problems, give them eight or nine to choose from. If assessing with tasks, either give multiple choices or, within a task, have the ability to personalize.

Diana's Corner

Providing choices allows me to see what my students know, instead of testing for what they don't know.

OF COURSE

Google Docs

In the directions for an assessment in Google Docs, tell students to clearly mark the selection of questions they would like to be assessed on.

Explore More

Sample Math Assessment in Google Docs:

goo.gl/bkEBjQ

Choose 5 of the 10 questions to respond to. Mark answers on the Google Form. Provide justification in the Google Doc.

Question 1

Which of the following sequences is arithmetic with the 100th term being 5020?

Justification for Answer

A. 1, 5, 9, 13, ...		
B. 70, 120, 170, ...		
C. 1, 3, 9, ...		
D. 1060, 2000, 2040, ...		

Google Forms

Instead of making all of the questions "required" in a Google Form, provide an option for students to opt out of some of the questions. For multiple choice questions, provide an option of "Did not choose this question."

Question 3

Cookies are sold individually or in packages of 2 or 6. How many ways can you buy 10 cookies?   2 points

- ☐ 6 ways
- ☐ 9 ways
- ☐ 12 ways
- ☐ 15 ways
- ☑ Did NOT choose this question.

Create sections in Google Forms to isolate the question choices. "Choose one question from each section" can be a way to ensure students address each learning objective.

> ### Explore More
>
> Sample Google Form with Multiple-Choice Response:
> **goo.gl/b1EWQD**
>
> Google Form Template—Choose 5 out of 10 Multiple Choice:
> **alicekeeler.com/5of10form**

Chapter 1-2 MC

Choose your questions

Question Menu
Which questions are you choosing? Reminder you are only choosing 5 of the 10 and you need to provide evidence on your Google Doc.

○ Question 1

○ Question 2

○ Question 3

○ Question 4

○ Question 5

○ Question 6

○ Question 7

○ Question 8

○ Question 9

○ Question 10

○ I am ready to submit

45. Provide Constraints, not Math Problems

Starting a lesson with a traditional math problem can lack context for student learning. Instead, begin your lessons with a story or problem to solve that helps students answer the commonly asked question: "why are we learning this?" Likewise, traditional word problems can be too specific, providing students with the exact information they need, and ultimately reducing critical thinking. A better approach is to pose a challenge. In other words, provide students with constraints for solving the problem or ask students to identify the natural constraints.

Think beyond the classroom when introducing constraints. Identify a big idea and have students create questions to investigate. For example, design a tank for a large, regional aquarium where fish or other aquatic animals can recover from illness or be held in containment before being

> ### Diana's Corner
>
> The best activities come from places that students know and visit. Think about regional aquariums, parks, zoos, and local farms or ranches. Each place abounds with opportunities when looking at constraints. For ranches, you need to know how much land is needed to support each animal (great for elementary level; you can add in costs when you get to the secondary level). The constraints can come from you as you introduce the process, or (better still) from research that is focused by the guiding questions that the students create.
>
>

released. The constraints come from the speed of the fish, their turning radius and jumping abilities, as well as the budget, the available space, and the space needed by the animals.

As students think more about the problem, they begin to personalize the activity. Demonstrating their thought processes and additional questions is part of their final product. This process allows you to really see the depth of understanding for each student; students cannot hide their understanding.

Production of new questions

These were the new questions that were generated as I thought more in depth about the question.
1.   How big is the tuna in question?
2.   Is the turning radius based on the size of the fish?
3.   The fact that the tuna is a schooling fish, does that influence the turning radius?

Activating Prior Knowledge

While reading the question, I began to activate prior knowledge based on my personal experiences of what I knew of each subject. The thoughts and ideas were scattered all over the place and are not all coherently connected. These images represent a rough representation of the ideas that were manifested inside my head.

Google Slides

Google Slides allows students to demonstrate their solution with videos, pictures, and drawings. In math, using Slides helps students to be concise. Students slow down and think about what they are putting on the slides. This practice helps with the ability to summarize and to take ownership of their work as they put information into their own words.

Explore More

Directions for Setting Up Projects and Slides Conversations:
goo.gl/x9Tbr8

Cooperative Learning Activity Slides:
goo.gl/iB5WHr

Cooperative Learning Activity

As you get ready to work:

- Create a list of essential questions that you will need to find the answers to in order to complete the task/activity that you are about to work on.
- Your questions will become the basis of your process and guide you through the work with a purpose beyond a single solution.

Monterey Bay Aquarium

 This activity is an activity based on an actual situation, with teacher creative license

Your group will design a backup aquarium to fit given constraints and then create a group Google Slide presentation of your process, the mathematics used and your concluding design

New aquarium needs to be created for the Outer Bay Exhibit's residents with the following constraints:
- Hold 1,000,000 gallons of water
- Needs to hold Yellowfin Tuna
  - take into consideration the tuna's turning radius
  - side must be a minimum of 2 feet above the water level because of the tuna's speed
- Will also hold sharks and turtles
- Aquarists will have to be able to walk around it to work/feed the animals

46. Play Ball

Playing games together as a class helps build culture and cultivates conversations around mathematical thinking. Creating an activity after something familiar to the students, such as baseball, helps to engage students in the activity. In class baseball, the students divide up into two teams and compete to answer math questions.

- Rearrange the room to simulate a baseball field. Move all the desks to the side of the room and line up the chairs into two benches.
- Each team needs a team captain to be the official spokesperson for the team.
- The first team sends up a "batter" who chooses a math problem.
- The other team works together to "catch the ball" indicating if the batter has a correct solution. If the batter is incorrect, switch teams."

Google Slides

Use Google Slides to create a game board that contains the questions. Use hyperlinks to link to other Slides and to link back to the main menu.

Explore More

Google Slides class baseball template:
alicekeeler.com/classbaseball

Directions document: **goo.gl/mtudQp**

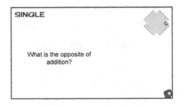

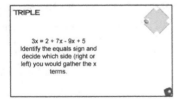

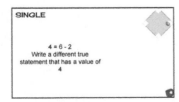

47. Digitize Analog Work

Glitter, scissors, and glue should not be abandoned. Sometimes technology is not the best tool. While work can be created on paper, whiteboards, or in dioramas, the work can still be submitted digitally. Google Docs, Slides, and Drawings all allow students to insert an image via "Take a snapshot." The webcam or document camera can be used to capture student work in the document or slide. Students can also use the Google Slides mobile app to add the image to the Google Slides app.

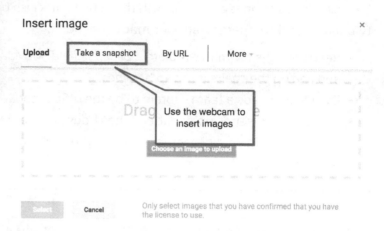

Submitting analog work digitally allows the work to easily be located later. And, like you, students can use the commenting features of Google Slides and Docs to provide feedback.

Google Slides

A few ways using Google Slides enhances the non-digital activity:

- Students can explain their activity.
- It's collaborative.
- Peer critique possibilities are improved.
- Students can embed a rotating slide-show into a website.
- Additional slides can be added to respond to feedback.
- It enhances persevering in problem-solving.

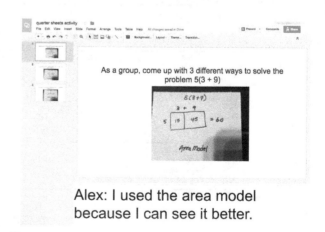

Alex: I used the area model because I can see it better.

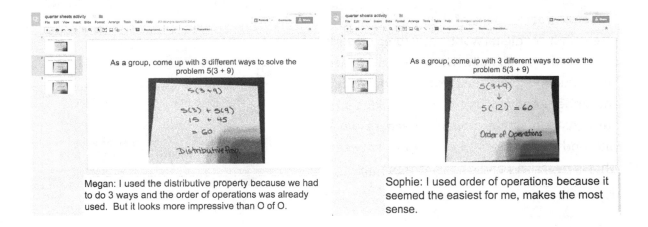

Megan: I used the distributive property because we had to do 3 ways and the order of operations was already used. But it looks more impressive than O of O.

Sophie: I used order of operations because it seemed the easiest for me, makes the most sense.

48. Discovery Activities

Learning in isolation is not as effective as learning collaboratively. Encourage collaboration; it allows students to make connections and to communicate effectively. Collaboration occurs within class and outside of class, and technology will allow for timely collaboration. With collaboration, all students benefit. All students get stronger in their understanding and are able to move to different DOK levels.

At the beginning or end of a unit, create Discovery Activities/Stations. Students will work with activities that

Diana's Corner

Discovery activities level the learning field. In my class, students come in with varied experiences. A set of discovery activities allows for students to experience the type of thinking that they are going to learn in the next section or unit. Then you can have a conversation to tie back to the activities. My students' favorite activity was a limit discovery lab where they used hair dryers, ping pong balls, Euler discs, whomper blasts, and more engaging activities with limits.

require more than one person to do easily. In the beginning of a unit, this activity will give a level playing field for students when they learn the math later in the unit. It will give everyone a common experience to reflect on. At the end of a unit, it allows for the work to be completed, summarized, and reflected upon.

Google Slides

Students create videos of their investigations, screenshots of digital work, or pictures of work completed non-digitally. Using Google Slides helps students build their levels of knowledge and understanding. The individual slides create constraints that students need to work with and allows them to distinguish between what is necessary and what is extraneous.

49. Start with a Picture

Mathematics does not simply exist in math books as equations. Help students see the math all around them by starting a lesson with the examination of a photograph. Have students look beyond the picture and think about what is happening in the photo. The idea behind starting with a picture is to encourage students to look at objects from different perspectives. Looking at objects through the eyes of a scientist or a mathematician is usually different than looking at it through the eyes of a historian or poet.

- Is there a constant rate of change (slope)?
- What types of motion are there?
- Is there a depth of field?

- What shapes are present?
- What repeated patterns are present?
- Categorize the objects in the image.
- Where are parallel lines represented?
- What geometry is represented?
- Describe fractions or proportions within the image.
- What objects are taller or shorter?
- What dilations can be identified?
- Examine textures within the image.

Google Slides

Inserting a picture into Google Slides can be accomplished by simply dragging the image onto the slide. Using a mobile device with the Google Slides app, the plus icon at the top allows for adding pictures directly to a slide from the camera. Use the Insert menu to insert an image. Use the Search option to locate ready-made pictures.

Explore More

Google Slides with Pictures Sample:
goo.gl/XVfgPy

Math Vocab Scavenger Hunt Template:
alicekeeler.com/ vocabscavengerhunt

Sample Scavenger Hunt:
goo.gl/rkikFx

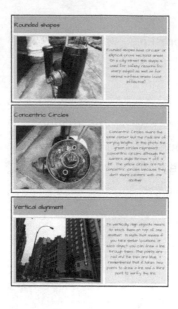

50. See Math Everywhere

Math teachers see math everywhere, but are students as observant of the math around them?

Bringing in real photos and fostering curiosity in students is a way to encourage students to start seeing math in the world around them. The following picture is of a Monarch Butterfly cluster. Ask students to estimate the amount of butterflies in the cluster.

Denis Sheeran provided this prompt for students to ponder from the image: *I love coffee. My wife says that if I ever die, they may actually find some blood in my caffeine stream. I'm interested in getting the most bang for my caffeinated buck. What should I buy?*

During the process of debating the solution, students realized the ice in the iced coffee made a difference, which brought all new variables and assumptions into the problem.

Google Slides

An easy way to "see math" is to use the Google Slides app. Here's the assignment:

- Create a blank Google Slide from the app.
- Insert a blank slide after the title slide and take a picture of something you see in your life.
- Insert a title and body slide to tell the story of your picture and to pose a mathematical pondering.

Explore More

Monarch Butterfly Class Discussion
Activity: **goo.gl/jZWYsk**

"I See Math" Project: **goo.gl/5Vpa4s**

Collaborative Describe the Math
Activity: **goo.gl/V7DJak**

Alice's Corner

Keep it simple. Leave the slides white. Not getting fancy makes it easier to create "I See Math" challenges for your students.

Conclusion

Using Google Apps to teach math is not about going paperless. Technology avails opportunities to better interact with students, collaborate, connect learning to authentic contexts, and increase critical thinking. If using technology is going to impact student learning, the task has to change. We hope the activities in this book have given you a plethora of ideas to try in your math class.

Share how you use Google Apps with math by using the hashtag #GoogleMath on Twitter. We are all better together! If you would like to allow others to use Google Apps resources you create, you can share them in alicekeeler.com/mathfolder.

> For additional ideas and resources join the Google Math newsletter:
> **alicekeeler.com/ googlemathnewsletter**

Google Tutorials

Google Apps

Google Apps is a suite of products, generally designed around collaboration. You can find most Google products by typing the name of the product into the browser followed by .google. com. For example, Google Drive can be located at drive.google.com. Google Drive provides online storage and allows users to create text documents (Google Docs), presentations (Google Slides), spreadsheets (Google Sheets), drawings (Google Drawings), and forms (Google Forms). These documents are saved within Google Drive for easy access across devices. Google Drive, Sheets, Slides, and Docs can be accessed through a mobile app in addition to the browser. The term "Google Docs" can refer to Google text documents, but is also used to refer more generically to the entire set of digital documents: Docs, Sheets, Slides, Drawings, and Forms. Using Google Apps eliminates your need to use Microsoft Office.

1. Creating a Google Doc

The term Google Docs can refer to Google Text Documents or the full suite of Docs, Sheets, Slides, Drawings, and Forms.

Google Docs, Sheets, Slides, Drawings, and Forms can be created by going to Google Drive and pressing the "New" button on the left hand side.

Alternatively, each of the types of Google Docs can be created by going to their individual website:

Docs.google.com   Sheets.google.com   Slides.google.com

Forms.google.com   Drawings.google.com

Using the app-specific website creates the document in Google Drive. Templates for the documents are also available on the app-specific websites.

2. One Version

A paradigm shift when moving to Google Apps is that you only need one version of a document. No longer do you need to make copies and backups. Using Google Docs you can be confident that you always have the correct version. Simply update the file and it is updated everywhere. The link to the syllabus on the classroom website does not need to be changed every year. Simply update the syllabus document to reflect the new school year. Updates to tests and department documents no longer need to be emailed with a request to replace the previous version. Eliminate version confusion and simply edit the test. Without any notifications, all the teachers have the correct version.

3. No Save Button

Something that can be hard to get used to is the lack of a Save button in Google Apps. The documents automatically save to Google Drive. Go to Google Drive and click on "Recent" on the left-hand side. Your documents are there even if you forgot to name them.

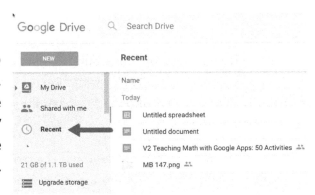

4. Using the Explore Pane

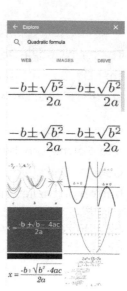

In Google text documents and Google Slides students are able to perform a search right in the document. Under the Tools menu, students select "Explore" or click on the Explore icon in the bottom-right corner of the screen.

Images and information from the search can be dragged directly into the Google Doc or Google Slides.

Google Drive

5. Sharing Files

By default, files created in Google Drive are private. From individual Google Docs, click on the blue "Share" button in the upper-right to add collaborators or change the sharing settings to allow anyone to view the document.

In Google Drive, single click on a file to reveal options in the toolbar. The link icon allows for sharing the file generically. The person icon next to the link icon allows you to share explicitly with certain people.

6. Share a Google Folder

Rather than sharing files one at a time, instead share a folder in Google Drive. Any files within the folder take on the sharing permissions of the folder.

7. Revision History

In the File menu of Google text documents, Slides, Sheets, and Drawings, you are able to view all of the revisions of a document. By choosing "Restore this revision" you are able to see who made what edits when and restore an old version.

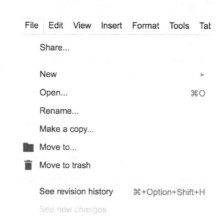

To view last year's syllabus content, use the revision history to view what the document looked like previously.

Choosing "Restore this revision" will update the file to look as it did at that point in time. No information is lost. You are able to go back to the most recent revision by restoring that revision.

8. Offline

Google Docs have some offline capabilities. Offline must be enabled in the settings of Google Drive. Students can create blank Docs, Sheets, and Slides in Google Drive when offline. Synced Google Drive files can be edited when offline. Features such as image search that require Internet connection are not available when offline.

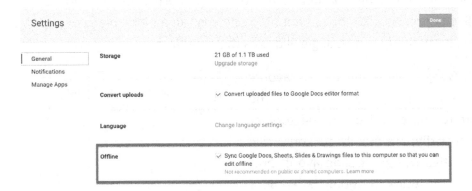

9. Converting Office Documents

The Chrome extension "Office Editing for Docs, Sheets, & Slides" (goo.gl/8WTEWO) allows Google Apps users to view and edit, with limited capabilities, Office documents in the Chrome browser.

Office documents can be converted to Google Docs by right clicking on the file in Google Drive and choosing "Open with." Math files oftentimes do not have the best conversion results. Moving to twenty-first-century skills for teaching math requires modifications to files; updating converted files for interactivity and searching questions is recommended. Transition to creating documents in Google Apps instead of using Office.

Google Text Docs

10. Make a Copy

Sharing Google Docs with view access does not allow others to edit or personalize the document. Use the File menu and choose "Make a copy" to obtain your own copy of a document.

11. Voice Typing

Voice typing can be activated by using the Tools menu and selecting "Voice typing." The keyboard shortcut Control+Shift+S can also turn voice typing on and off. Additionally, with practice, students are able to edit their Google Text documents using voice commands. For more information visit the Google Help page (goo.gl/NmIUMt).

Google Slides

12. Edit the Slide Master

Use Google Slides to get information instead of giving information. Students can interact with a Google Slides presentation document. Each slide is like a blank piece of paper, so students can do almost anything using Google Slides as their medium.

By default, Google Slides has eleven layouts when inserting a new slide. Clicking on the arrow next to the plus icon in the toolbar reveals layouts that are not always conducive to math activities. Designing slide layouts provides the opportunity to guide a student activity with helpful graphic organizers.

The slide master is located under the View menu (choose "Master"). Edits made on the slide master are universal. Making any font or color changes in the slide master affect all slides utilizing that layout.

The eleven default layouts can be edited or deleted. Layouts that are in use are not able to be deleted. The first layout is in use by default. Slides layouts, two through eleven are generally available to delete. New layouts can be added using the plus icon in the toolbar.

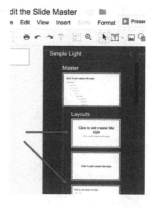

The "Master," located above the layouts, dictates global settings for all layouts. A graphic or text box added to the "Master" will appear on all Slides layouts and can only be removed by removing them from the "Master."

Elements created on the slide layout are locked down to the slide. When students use a slide layout in the presentation, any images or text boxes are not editable or moveable. Text placeholders provide a text box that students can fill out.

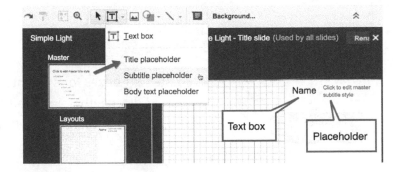

In the example (goo.gl/QRenmC), only one slide layout option is available to students. Graph paper was added to the layout. Students cannot edit or move the graph paper. The word "Name" was added to the slide; students are unable to edit or move the text. A place for students to enter their name is available. While students can easily add their own text boxes to a slide, placeholders are helpful to indicate where text is expected.

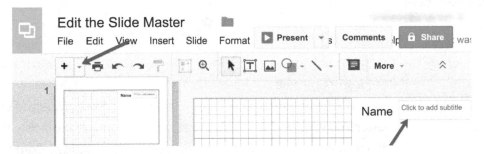

13. Create Movable Objects

Objects created on the slide master are locked down. Movable objects must be created on the slides in the filmstrip of the Slides presentation. We suggest creating these objects "off canvas."

1. Zoom out on the slides to reveal more of the "off canvas" area.

2. Draw a shape or text box, or insert an image.

3. Click on the object and use CTRL-D to duplicate the object multiple times.

4. Select the group of duplicated objects.

5. Use the Arrange menu to "Align horizontally" in the "Center" and to "Align vertically" in the "Middle." This creates a stack of the object that can be dragged onto the slide canvas.

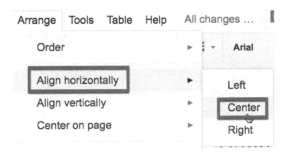

Google Sheets

14. Merge and Wrap Cells in Google Sheets

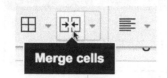

To create text boxes for students to respond in, it is recommended to merge cells together and set the text wrap. Highlight the range of cells and use the "Merge cells" icon in the toolbar.

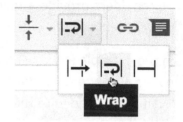

To the right of the "Merge cells" icon are the icons to center the text horizontally and vertically. The Word Wrap icon is by default selected as "Clip text." Use the small arrow on the word wrap icon to choose the middle option of "Wrap."

15. Conditional Formatting in Sheets

Conditional formatting allows for cells to be filled with a particular color in response to the content within the cell. This allows for creating self-check spreadsheets or for analyzing student responses from a Google Form. Highlight the range of cells you wish to apply the conditional formatting to. Choose "Conditional formatting" from the menu. A sidebar appears to allow you to set the rules for the selected range.

16. Cell Referencing in Sheets

When writing spreadsheet formulas, we suggest using cell referencing, which uses the cell address rather than a numerical value. For example if cell A5 contains the value 12 and cell B5 contains the value 4, the formula =A5/B5 will return the value of 3 based on the cell reference. Changing the values in the cells results in the formula automatically updating the result.

Copying and pasting a formula will adjust the cell reference. If the formula =A5/B5 is copied and then pasted into the cell below it, that formula will adjust to =A6/B6. This is often desirable.

When using constant values, referencing the same cell value repeatedly, Absolute cell referencing indicates that the formula will always reference a particular cell. Use the dollar sign symbol to lock down the reference to the row and/or column of a cell. =A5 will always reference cell A5. =$A5 will always reference column A, but not necessarily row 5.

fx | =(D5-B5)/(C5-A5)

	A	B	C	D	E	
1	x1	y1	x2	y2	slope	Interpret each slope.
2	-52	-100	79	-11	0.679389	
3	25	44	46	-68	-5.33333:	
4	56	-66	-45	-29	E5 -40 × 3(	
5	-51	80	-48	-40	=(D5-B5)/(C5-A5)	
6	41	-12	-86	-54	0.330708	
7	-92	29	-14	98	0.884615	
8	17	18	49	11	-0.21875	
9	61	-41	18	-6	-0.81395:	

17. Concatenate Text and Values

Occasionally, combining text with values from the spreadsheet in the same cell helps to more clearly communicate what is happening in the spreadsheet. For example, when students calculate the slope in a spreadsheet they may use a formula such as =(U6-U11)/(X17-AB17). The resulting value would be a lone number in the spreadsheet.

Values can be concatenated (put together) by using the ampersand (&) symbol. Text values need to be enclosed with quotations. Be aware of spaces, because they need to be manually entered within the text quotes.

="Rise= "&U6-U11

="Run= "&X17-AB17

="Slope= "&(U6-U11)/(X17-AB17)

="Slope= "&(U6-U11)/(X17-AB17)

Google Forms

18. Keyboard Shortcuts in Forms

Use the keyboard shortcut Control+/ to view a list of keyboard shortcuts available in Google Forms. Create a new question with the keyboard shortcut Control+Shift+Enter.

Google Classroom

19. Creating Assignments in Google Classroom

In a Google Classroom class, click on the Plus icon in the bottom right to add an assignment to the Stream.

1. Post to multiple classes with the drop-down arrow.

2. Title the assignment.

3. Describe the task.

4. Choose due date and time.

5. Choose assignment topic.

6. Attach files from Google Drive.

7. Change attachment settings:

    Students can view file
    Students can edit file
    Make a copy for each student

8. Assign the assignment, save it as a draft, or schedule it to post at a specified time.

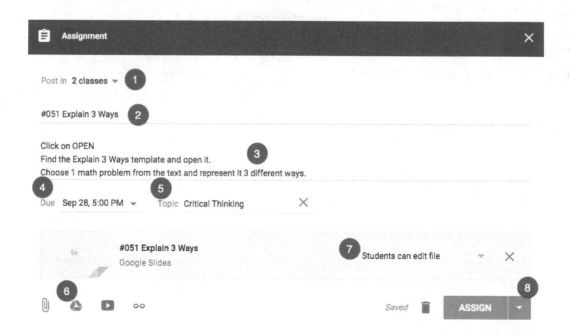

20. Attaching Files into Google Classroom

By default, Google Docs attached to Google Classroom are "Students can view file." Clicking on the arrow allows you to change the file to "Students can edit file." This allows all the students in the class to edit the same document. The third option, "Make a copy for each student," creates an individual copy of the file per student.

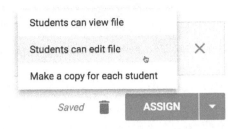

21. Attach Google Forms in Google Classroom

Attach Google Forms in Google Classroom by using the Drive icon when creating an assignment. Students are automatically linked to the live view of the Form. Teachers can access the spreadsheet of results right from Google Classroom. Students do not have the link to the results. Teachers can view responses in the spreadsheet as they are submitted.

When Google Forms are the only attachment on a Google Classroom assignment, submitting the Form automatically marks the assignment as *done* for the student. When additional things are attached the student must manually go back to the assignment to *Mark As Done* or to *Turn In* the assignment.

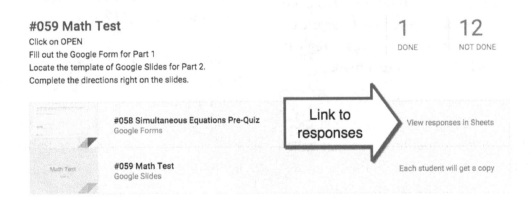

Add-Ons

Add-ons are third-party (not made by Google) scripts that give additional functionality to a Google Doc. Add-ons are available in Google text documents, Google Sheets, and Google Forms. Look for the Add-on menu in the App to add.

Google Docs Add-ons

g(math) (goo.gl/i6E8ga)

g(math) allows students to insert sketches of math, graphs, and math symbols right into a Google Doc.

gTable Calculator (goo.gl/6rOnCu)

Math calculations are not just for spreadsheet tables. This Add-on will calculate math on a series of cells in a Google Doc table. The Add-on also provides a regular calculator right in the side-bar menu.

WizKids CAS (goo.gl/bWkBoS)

From the sidebar of a Google Text Doc, WizKids solves equations, plots graphs, finds solutions, simplifies expressions, factors expressions, and inserts results into the Google Doc.

Gliffy Diagrams (goo.gl/qtyU11)

Students can create diagrams to compare and contrast the solutions to math problems.

Lucid Charts (goo.gl/mqBj7Y)

Another diagram tool to help students clearly communicate their ideas.

Table Formatter (goo.gl/L1pRWs)

Tables can be an effective way to communicate information. Displaying information in a visually pleasing way is an excellent skill for students inside of math class and out.

Formula Editor (goo.gl/ITzVjX)

Creates a sidebar to allow students to create and insert math symbols into their Google Doc.

Auto-Latex Equations (goo.gl/UUSmoH)

Teach students to type their math equations in LaTeX. This Add-on will render the equations in the document from LaTeX into an attractive equation or expression.

Brainstorming Race (goo.gl/uc7H3f)

Have students brainstorm applications of math concepts collaboratively. Then, the Brainstorming Race Add-on shuts down editing and creates a Google Form for collaborators to vote on.

Google Sheets Add-ons

Autocrat (goo.gl/wuPMUU)

Autocrat Add-on merges data from a Google Form to a formatted Google Doc or Google Sheet. Note that this is a Sheets Add-on for the Spreadsheet created from a Google Form.

Statistics (goo.gl/hRxx7B)

The Statistics Add-on does statistical data analysis and regression modeling for Google spreadsheets.

Doctopus (goo.gl/Yw2wDb)

Doctopus copies templates of documents for students and allows the teacher to share and monitor docs. A folder structure for each student is created in Google Drive.

Flubaroo (goo.gl/DraIHr)

The Flubaroo Add-on grades Google Form responses when used as a formative assessment quiz. Form responses can be graded immediately upon Form submission or the teacher can initiate Flubaroo grading the responses. Add the Add-on to the spreadsheet connected with the Google Form.

Autocrat (goo.gl/BdAhX0)

The Autocrat Add-on merges Form data to a Google text document or Google Sheet. Autocrat can be an effective way to send feedback to students. Create a feedback column in the Form responses spreadsheet. Use Autocrat to send the student their responses and your feedback. Autocrat can also be used to send students mastery certificates. Add the Add-on to the spreadsheet connected with the Google Form.

Copy Down (goo.gl/3EHGOs)

Under the Add-on menu, the Copy Down Add-on will automatically apply the =image() formula to form submissions. Add the formula to row 2 in the spreadsheet destination for the Form. Copy Down will automatically apply the formula to all form submissions.

rowCall (goo.gl/ne8P5x)

Create a sheet for each student's data from Google Form data. This Add-on takes a piece of data that repeats and filters the Form data to segregate each unique entry and filter the responses that go with that piece of data.

Google Forms Add-ons

g(Math) for Forms (goo.gl/NIz5ms)

g(math) is an Add-on that allows teachers to insert math symbols or graphs into a Google Form. Teachers are able to enable student responses with math symbols.

formLimiter (goo.gl/hFohY8)

formLimiter shuts off a Google Form after a max number of responses, at a date and time, or when a spreadsheet cell equals a value.

Chrome Extensions

Chrome extensions add additional functionality to the Google Chrome browser. Some examples of extensions are screenshot tools, tab counts, ad blockers, color pickers and more. Visit the Chrome webstore (chrome.google.com/webstore) to add extensions.

Drive20 (alicekeeler.com/drive20)

The Drive20 Chrome extension opens up 20 student documents at once from Google Drive. This makes the feedback process faster.

Alice Keeler QuickShare Screenshot (goo.gl/D5BHwG)

While many Chrome extensions will take a screenshot, QuickShare only shares to your Google Drive and automatically copies the link to the image.

Alice Keeler Open Side by Side (goo.gl/d2bHTA)

After installing this Chrome extension, right-click on a link and choose to "Open Side by Side." This opens the link on the right side of the screen and moves the current window to the left side.

Side-by-side windows can be helpful when creating resources or when assessing student work. When transferring information from a Google Doc or other format to a Google Form, having both documents open side by side makes the transfer process easier. Having Google Classroom or the gradebook open side by side with student work makes feedback comments easier. Students can take advantage of side-by-side windows by having the directions side by side with their work.

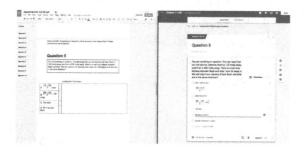

Tab Resize (goo.gl/mJYCQ)

Allows for multiple browser windows to be tiled in a variety of formats.

Tab Scissors (goo.gl/2p44o)

When multiple tabs are open, Tab Scissors quickly separates the tabs and displays them side by side.

Tab Glue (goo.gl/zVfsu)

Reversing the splitting of windows, Tab Glue merges the windows back into one. Each window has its own tab.

Alice Keeler Classroom Split (goo.gl/hdvvPx)

Students are able to use Classroom Split to open their assignments list in Google Classroom side by side with their work.

Alice Keeler Gradebook Split (goo.gl/UI4wKv)

Add the URL to your online gradebook into the extension settings to open your gradebook side by side with student work.

OneTab (goo.gl/w8iWVL)

When using web resources, the number of open tabs can cause your computer to get sluggish. Use OneTab to close the tabs in your browser and create a list of the links to return to later. Search for a collection of web resources on a particular topic and use OneTab to share the list of websites.

Additional Resources

Explore More
Google Drive Folder of Math Resources:
goo.gl/5Hg7uf

Protractors for Google Slides
This Google Slides template provides several transparent protractors. Adding images to a slide will cover the protractor. Use the keyboard shortcut Control+Shift+Down-Arrow to send the image to the back, behind the protractor. The protractors can be resized or rotated. The template is in the Google Drive resources folder at alicekeeler.com/protractors.

Google My Maps Resources by Donnie Piercey
Donnie Piercey, Google Certified Innovator, has curated a resource on Google My Maps. Find ideas and resources for using Maps with students: **goo.gl/XcfRkz**.

Desmos
Desmos is a powerful tool to have students explore math concepts. View and use the resources available at teacher.desmos.com/

Tootsie Pop Statistics Investigation, by Denis Sheeran: **goo.gl/hV6j45**

Prime Numbers Polygraph, by Denis Sheeran: **goo.gl/LnK2EL**

Gateway To Quadratics, by Denis Sheeran: **goo.gl/xIR5K2**

Integer Game, by Andrew Stadel: **goo.gl/pjQYVb**

Waterline, by Desmos: **teacher.desmos.com/waterline**

Function Carnival, by Desmos teacher: **desmos.com/carnival**

Symmetry, by Suzanne von Oy: **goo.gl/2szycO**

Desmos Scientific Calculator
Access this scientific calculator at: **desmos.com/scientific**

Daum Equation Editor
An equation editor allows you to easily create formulas of any length and edit them. If there is a need for having an equation, this equation editor will give you a professional look for your documents.

Daum Equation Editor: **goo.gl/xvoK2**

Instant Relevance

Denis Sheeran (@MathDenisNJ), math teacher and author of the book *Instant Relevance* (available on Amazon), shares on his blog (Denissheeran.com) math activities and ideas that incorporate Google Apps for Education into the math classroom across elementary, middle, and secondary grade bands. Denis is a Google Certified Trainer who also speaks nationwide on how to incorporate Google Apps into the mathematics classroom by using the SAMR model to reach full integration. Some recent posts by Denis include:

"Teaching Math with Google Apps, a 4 in 1 Approach":
denissheeran.com/googlemathapps

"Top Google Sheets Add-ons":
denissheeran.com/sheetsaddons

"Geoguessr Hyperdoc Activity":
goo.gl/eemJJp

YouCubed

The mission of YouCubed is "Inspiring Math Success for all Students through Growth Mindsets and Innovative Teaching." Visit the website for math activities, research-based teaching strategies, and an online course for teachers. www.youcubed.org. This site will give you a strong foundation to grow as a teacher and help your students learn at a deeper level.

The Classroom Chef

Throughout the book *The Classroom Chef*, John Stevens and Matt Vaudrey share the struggles and strife of being in a stale classroom and how they found inspiration to take a risk and built a culture of curiosity. The examples throughout their book combine a splash of creativity with a heavy dose of curiosity in an effort to make learning—particularly math—meaningful. Follow them on Twitter at @classroomchef and check out **classroomchef.com** for more information.

The Classroom Chef: Sharpen Your Lessons, Season Your Classes, Make Math Meaningful is available on Amazon.
classroomchef.com/book

Open Middle

Going beyond the predominantly DOK 1 math problems in the text can be challenging. Open Middle provides examples of DOK 2 and DOK 3 math problems.
openmiddle.com

3 Act Math

Dan Meyer provides free math curriculum on his blog **blog.mrmeyer.com**. 3 Act math problems present students with a situation in Act 1, usually in picture or video form, that can create a variety of questions. The teacher knows which question to follow, then reveals information that can help the students progress in the problem (Act 2). In Act 3, the solution is revealed, usually by completing the video or displaying more of the picture. Students then engage in a follow-up extension based on some of their other questions. Dan provides an explanation on his blog post **blog.mrmeyer.com/2011/the-three-acts-of-a-mathematical-story**. In the sidebar of the site, a Google Sheets resource of 3 Act math activities is provided.

Matific

Matific provides digital games that build conceptual understanding. Matific allows for differentiated practice along with data for the teacher to help guide learning and set learning goals. **matific.com**

Wild Maths

Wild Maths understands that mathematics is a creative subject and provides games, activities and stories to help students think creatively about mathematics. **wild.maths.org**

Socially Conscious Mathematics

sociallyconsciousmath.wordpress.com This website is a source for relevant essential questions that go beyond DOK 1 thinking.

Virtual Manipulatives

Virtual manipulatives allow students to explore their learning in class and out. Google Chrome Apps provide students with opportunities to interact with math concepts.

Number Pieces: **goo.gl/UVyyS5**

Number Frames, by The Math Learning Center: **goo.gl/XuQUbF**

Geogebra Math Apps: **goo.gl/H28nq**

Robocompass: **goo.gl/oXlVI9**

Mathigon—World of Mathematics: **goo.gl/tf2kIL**

Lego Builder: **goo.gl/LikBJ**

Number Rack, by The Math Learning Center: **goo.gl/zXvHVs**

Number Line, by The Math Learning Center: **goo.gl/XhY4Ep**

Pattern Shapes, by The Math Learning Center: **goo.gl/EGIao6**

Fraction Wall, by Visnos: **goo.gl/wfBnxt**

Fraction Puzzle Game: **goo.gl/nhrU9Z**

Number Balance: **goo.gl/A2CLRX**

Tynker

The Chrome App Tynker (**goo.gl/2I1wMh**) teaches students to code. They are able to create their own math games or activities to demonstrate their understanding of math. Tynker integrates with Google Classroom to allow teachers to easily track and provide feedback on student projects.

Diana's Story

Change is hard, but doing nothing puts your students further behind. Galileo is credited with stating, "Mathematics is the language ... [of] the universe." As teachers, we need to let math speak so students can listen to and learn from it instead of reaching for their earbuds. Allow students to see the beauty of mathematics in the world around them. Our hope is to share with you options to help open the world of mathematics to your students through the use of technology, an engaging tool for learning.

Students who are in charge of their learning, engaging in conversations, and debating mathematics are a wonder to experience. What kids get out of a class when they are active in the learning process sets them on a successful path for their future. As a teacher, you should be a life-long learner, so take the challenge and change one thing this next semester to begin your technological journey of student engagement; after all, math is not a passive subject. The technology that we have at our fingertips today should be embraced and utilized to its fullest potential in the hands of our students.

Why is there a need for change? Look at your class—there in front of you is your reason. As a middle school and then high school teacher I used to think having students show their steps and reflect on what they did was enough to show their understanding. Then Wolfram Apps made me rethink what level of questions I was really asking. Most of them could be answered with apps, step by step, with all work shown. I had to move away from just skill-based questioning and have students compare and contrast, recognize patterns, and hunt for generalizations and connections with what they have done in the past. It opened up a new world for me and my students.

My classroom has always been 36+ students. Homework was not my favorite thing—in fact, I am not sure why it should even be assigned. So my "homework" has become conversation and collaborative learning time. During class, we are investigating and working hard at letting students be the center of the classroom (hard for students, since they have mostly experienced teacher-centered classrooms with lectures and possibly online homework). They come into my classroom and immediately panic and put up their barriers, but slowly they see the power of what we are doing and end up engaged and enjoying the class (reality check time—it takes about six weeks for the transformation to occur). Most importantly for me, their dislike for math becomes enjoyment and fun with math. They begin seeing math everywhere. The use of technology and hands-on opportunities with and without technology has helped them change their attitudes. Technology has allowed for deeper understanding of the mathematical concepts we are working on. I have them choose three problems/activities to work on with me over a week, or longer if needed. The problems/activities are done in Google Slides where we can communicate with each other through comments. With no grade that I need to deliver, I get to ask questions or talk about what I see. Students respond back, and our conversation continues until we are all comfortable with their understanding. Students begin asking "What if … ?" questions, and actually investigate them on their own and share with the class some of the things they have noticed. The entire class benefits from their enthusiasm, and it begins with a conversation between two people collaborating and questioning.

My students need to do well on a state assessment. To help them retain their new knowledge I have them create a digital portfolio focusing on the mathematical domains for their state assessment. Staying focused on their purpose and providing reflections on the pieces entered, allows them to create a strong portfolio that they can use as their study guide for the rest of their academic journey.

Structuring a Math Lesson

When creating a math lesson, it's important to keep in mind the 4 C's, Mathematical Practices, the 5E Instructional Cycle, Polya's 4 Steps of Problem Solving, and the DOK levels. That sounds like a lot, but when you really look at each of these models or tools, incorporating them is simply good teaching practice. Here's the good news: If you choose one to start with, you will find that you automatically cover the rest. And together these tools will help you engage students in learning. In this section, we'll touch on the ways these tools can work together to help you engage students in meaningful math lessons.

Before we get into the explanation of each tool, we want to remind you that every math lesson needs a focal concept. *Everything* you do as part of that lesson should point in the direction of that content. Learning how to consistently acknowledge the focal concept is one of the most difficult things to do—it's easy to get sidetracked! But that sharp focus will help you plan powerful, effective lessons. Once you have your focal concept and begin structuring your lesson, create a flowchart through the content that identifies how everything—from the assignments to the assessment—is connected to that primary focus.

Depth of Knowledge

Number sense is the key to being good at math; not being fast.

Webb's Depth of Knowledge (DOK) levels focus on the amount of critical thinking in which a student is involved. DOK levels are difficult to assess because they are not defined by the activity, but rather by the critical-thinking levels. DOK is not about how difficult a task is, but rather how complex it is. Challenging math problems can still be a DOK level 1 if students are following a set of directions or an algorithm.

Today's math classes need to increase a focus on strengthening critical-thinking levels for students. The majority of current math textbooks focus on foundations and lower-level DOK questions. Teachers are the ones who take the thinking to the higher DOK levels. It is with strong questioning techniques that students move from skill level understanding and engage in critical thinking and problem solving. Although the move to higher DOK levels requires a teacher's nudge, one of the easiest ways to nudge students is to not tell them what to do. Instead, have students come up with their own guiding/essential questions. Asking the questions helps students develop their problem-solving skills while simultaneously raising their DOK levels. And, as just one example of how focusing on one model ensures that you will meet the others, when students ask their own guiding/essential questions, they enter Polya's 4-step problem-solving process.

For examples of math problems at different DOK levels, check out Robert Kaplinsky's website: **robertkaplinsky.com/tag/depth-of-knowledge-dok/**

DOK Level 1

This is the recall level. The focus is specific facts, specific details, definitions, algorithms, or specific procedures.

There is only one correct answer or a combination of correct answers.

Math problems ask students to follow a set of procedures: an algorithm.

Practice problems following examples in a textbook are typically DOK 1.

Most worksheets are at the DOK 1 level.

DOK Level 2

This is the skill/concept level. The focus is on applying skills and concepts.

Students are asked to explain how or why.

There is only one correct answer or combination of correct answers.

Students make a decision in calculating the math problem.

The problem the students are solving is significantly different from the examples.

DOK Level 3

This is the strategic-thinking level. The focus is on reasoning and planning.

The thinking level is complex and abstract.

At this level, students will be defending reasoning or conclusions with evidence.

There are multiple approaches and multiple answers.

DOK Level 4

This is the extended-thinking level.

The focus is on making real-world applications in new situations

The thinking level is about complex reasoning and planning.

Students answer open-ended questions.

Problems maybe have multiple answers or approaches and usually take an extended period of time to complete.

4 C's

The 4 C's are about preparing twenty-first-century students for a global economy. The following four specific skills were identified by the Partnership for 21st Century Learning (www.p21.org) to help students find success in school and in the real world.

Critical Thinking

- Students are able to solve different types of unfamiliar problems in traditional ways and innovative ways.
- Students are able to identify and ask significant questions to help them understand and help lead them to a strong solution.

Communicate

- Students can communicate ideas clearly. This is not necessarily a verbal exercise.
- Students are able to articulate their thoughts and ideas effectively in a variety of forms and contexts.
- Students are able to assess the impact of their communication.

Collaborate

- Students are able to work effectively and respectfully with one another.
- Students are flexible and willing to make necessary compromises to accomplish a common goal.
- Students share responsibility for their collaborative work and value the individual contributions.

Creativity

- Students try a variety of ideas.
- Students think of "failure" as the opportunity to learn and think differently and more creatively.
- Students look for multiple ways to solve a task and show innovation in how they approach a solution.

Mathematical Practices

The Mathematical Practices for the Common Core State Standards (CCSS) are the foundation for our students' ability to think mathematically and communicate mathematics successfully. Below we've identified the standards students must meet as well as what that might look like:

- **Make sense of problems and persevere in solving them**
  - Students are able to figure out what the problem is asking.
  - Students attack the problem until it is done.
  - Students are able to solve tough problems by applying what they already know.
  - Students self-assess their process and progress.
- **Reason abstractly and quantitatively**
  - Students will be able to deconstruct the problem.
  - Students work with problems in multiple forms (algorithms, graphs, diagrams, etc.).
- **Construct viable arguments and critique the reasoning of others**
  - Students will talk to one another using appropriate academic vocabulary.
  - Students will be able to support or give suggestions for improvement for their peers' work.
  - Students will understand that it is all about the work and not the individual when coming up with arguments and critiques.
- **Model with mathematics**
  - Students will be able to look at the world outside of their classroom and be able to create and solve practical problems.
- **Use appropriate tools strategically**
  - Students will be able to select and use the appropriate tool for the situation in front of them.
  - Students understand for themselves what tools they need to accomplish the task.
- **Attend to precision**
  - Students will use appropriate academic language.
  - Students will make a conscientious effort to include detailed explanations and precise solutions.
- **Look for and make use of structure**
  - Students will look for patterns and their repeated use for more complex problems.
  - Students will be able to deconstruct a problem/task and use the smaller structure to solve it.
- **Look for and express regularity in repeated reasoning**
  - Students will explore concepts and data to generalize the patterns that they notice, derive formulas, and make sense of observations.

5E Instructional Cycle

The 5E Instructional Cycle focuses on what the teacher can do to help get students to engage, explore, explain, elaborate, and evaluate. The cycle helps guide teachers through questioning ideas and practices to keep students motivated and moving forward. They are quality teaching practices that allow students to access the eight CCSS Mathematical Practices.

Engage

Engaging your students begins their learning process. As the teacher, you need to create interest, spark curiosity, and raise questions/problems to help elicit responses to uncover your students' current knowledge. The goal is to get them to want to ask questions such as, Why? What if … ? How can this problem be solved?

Explore

Exploring gives all students a common base of experiences for the lesson. As the teacher, you encourage your students to work together without direct instruction. As you observe and listen, consider what probing questions you can ask to redirect students. Make sure to provide plenty of time for students to puzzle through their lessons. For this instructional cycle you, the teacher, are simply a consultant for your students. (As a learning consultant, you are not passive; you are observing and listening.)

Explain

When students must explain their work, they learn to really focus on what they are doing. This is where students demonstrate and showcase their conceptual understanding and process skills. As the teacher, you encourage students to explain concepts and definitions in their own words, you ask for evidence and clarification, and you help students use previous experiences as the basis for explaining. This is also where you can provide definitions, explanations, and new vocabulary.

Elaborate

Elaborating challenges your students to extend their conceptual understanding and skills. It helps to create the opportunity for deeper and broader understanding. As a teacher, you will encourage students to apply what they have learned to new situations and have them look for alternative explanations. Your questioning techniques will allow for this to happen.

Evaluate

Evaluating is embedded within the other aspects in this instructional cycle. Students assess their understanding and abilities and provide opportunities for you to evaluate their progress as well. You will observe and look for evidence that students have changed their thinking. Ask the questions: What do you think? What evidence do you have? What do you know about the problem? How would you answer the question?

Read More

Read more about the 5E Instructional Cycle and the research behind it here:

goo.gl/RggF5B

Show Your Steps

A staple in the math classroom for a long time has been to "show your steps" or "show your work" when solving problems. This practice allows teachers to see that the students understood what they were doing. Technology is changing this idea. Google, Wolfram Alpha, and many other math apps and websites will not only calculate the answer to any math problem for students, but will also show all of the steps. All of the math skill problems traditionally asked of students can be solved, with all steps shown, using technology. We need to rethink what we are asking students.

Instead of trying to fight technology, we can use different questioning strategies. In a technological age, we can move past focusing on the procedural steps to deeper understanding and applications, which means students can work at higher DOK levels.

Working out math problems by hand has value, but sometimes students get so lost in "the steps" they miss the bigger picture once they have "the answer." For example, having students graph multiple quadratic graphs by hand in order to recognize patterns for what the A coefficient does can cause students to get lost in the plotting of points instead of the overall graph because students equate time spent on a task as the most important aspect of the assignment. Using a graphing app such as Desmos allows students to investigate and "get" what the A coefficient does in seconds by fitting a quadratic graph to an image using sliders. Focusing on the concept rather than the computation gives students deeper understanding. This also allows students to ask and answer "what if" in a matter of seconds. Likewise, a book assignment of completing thirty problems or a worksheet of similar problems can deny the students the ability to think deeper. If they are focused on the procedural steps (DOK 1) rather than focusing on understanding the concept, they miss the point of the lesson.

Embrace the technology that exists. Using Wolfram Alpha, or other programs, select two or three problems and have students compare and contrast the procedures they notice. Students can then take the pattern they noticed and work on another problem to test the procedures out themselves. Students can engage in a class/group discussion on the success of what they did, and now they have ownership of the new procedures for problem solving.

Technology allows us to spend our time doing mathematics differently. It gives us the opportunity for going deeper. Focusing on explanation and application becomes the class culture. When students wrestle over ideas rather than "learn" the steps, they take ownership of their learning and are proud of it.

Diana's Corner

Start Class in the Middle

A traditional math lesson follows a cookie-cutter format: go over homework, lecture with notes, do a few sample problems, have students do a few problems, close the lesson. In textbooks, it is the two-page spread. Unfortunately, this format prevents many students from seeing how everything is connected. One way to help change the classroom to a more connected learning opportunity is to adjust the beginning of a lesson to be halfway through the class period. This allows for the middle and end of the class to be basic foundational information/activity. When students return the next day, they explore the extensions of the concept and then bridge to the new concept. Instead of the class beginning "easy" and ending "hard," class begins "hard" and ends "easy" with a bridge to connect their learning.

Note that this format doesn't leave time for going over homework. That's okay! The concentration is on learning. Students do receive feedback through comments, so the need for going over the homework daily no longer exists.

Think about how many times your class has fallen behind in its schedule because too much class time was spent going over homework answers instead of covering the content.

This middle-first format sounds easy, but for the teacher it can be hard. The difficulty is breaking the habit of the lesson format. Notice what happens, though, when you change: students see the connection of the lessons, they leave class with the foundations of the next section and a curiosity of what will happen during the next class meeting. They have that important "sit time" with a concept before they start digging deeper. This format is a student-empowering learning opportunity. Traditional homework is not necessary. Instead, reflective questions can be used to challenge students to think, and that helps keep their curiosity alive. For example, review the three to five problems that you chose for the students to do with you during class, then the overnight question could be, "Compare the processes we used in class today to solve the problems. How are they the same and how are they different? Which of these processes reminds you of something else that we have done earlier this year, or in previous years?"

Student Ownership of Learning

Students need the opportunity to take ownership of their learning instead of being teacher-dependent. This doesn't mean that the teacher isn't necessary; to the contrary, this approach allows the teacher's role to become more important as an academic coach and the source for stronger DOK levels. Thought-provoking questions, projects, and engaging tasks all help students take ownership of their learning. It is exciting to see a student energized and wanting to share what they have learned, and the added bonus is that they *remember* it.

Students are surrounded by resources, and your classroom should embrace their opportunity to utilize them. Allow your role of information-giver to become that of a guide—guide students through the information, ask them questions that deepen their understanding, and help them understand the resources that are available to them. What determines how much and how well students learn is the dynamics of their classroom, the daily collaborative interaction with other students, and their teachers. As a teacher, you can create that dynamic in your classroom for your students.

One interesting benefit of student ownership of learning is that they are charged with creating their own guiding/essential questions to answer. When students have the opportunity to struggle with a problem, talk about the problem, and process what has been said, they are on their way to owning their knowledge. Imagine the teacher or a student presenting an interesting, researchable problem. The students struggle with it, talk about it, share ideas about what they think, collaborate on various solution methods, and then the teacher or a student summarizes the class's statements. Students will own that process and use it in every aspect of their education, whether in the classroom or outside of it.

Mathematical Mindset

by Jo Boaler

Using technology is not just about making what you are doing digital. It is a shift in mindset. Instead of asking the question, "How can I do this digitally?" ask, "How can I interact better with students?" or "How does this change how the classroom looks?" Digital tools allow us to spend more time talking with students rather than at students.

We highly recommend the book *Mathematical Mindsets* by Jo Boaler, a Stanford University math professor and researcher. Her book explains important paradigm shifts that are necessary for helping students to learn math, and more importantly, to love to learn math. Available on Amazon: goo.gl/mGSRNo

Technology is a tool that allows students to investigate mathematics in ways that prior generations couldn't. As the teacher, you will ask guiding questions to help students reach their potential and push their DOK levels. *Mathematical Mindsets* gives each teacher the support and knowledge to help themselves and their students continue investigating and enjoying their mathematical journey. The support the book gives allows both the teacher and the student to become successful risk takers.

Acknowledgments

We would like to thank the following people. Without their support, ideas, and feedback we would not have been able to complete this book:

Denis Sheeran (@mathdenisNJ)

Donnie Piercey (@mrpiercEy)

Ronessa Acquesta (@ronessaacquesta)

Chelsea McClellan (@marvelousmcc)

Cathy Cheo-Isaacs (@iwearthecrowns)

Mandi Tolen (@TTmomTT)

Jay Murphy (@mrteachnology)

Jonathan Rochelle (@jrochelle)

Howard Hua (@howie_hua)

Kendia Herrington (@ScienceKendia)

More from DAVE BURGESS Consulting, Inc.

Teach Like a PIRATE

Increase Student Engagement, Boost Your Creativity, and Transform Your Life as an Educator

By Dave Burgess (@BurgessDave)

Teach Like a PIRATE is the New York Times' best-selling book that has sparked a worldwide educational revolution. It is part inspirational manifesto that ignites passion for the profession, and part practical road map filled with dynamic strategies to dramatically increase student engagement. Translated into multiple languages, its message resonates with educators who want to design outrageously creative lessons and transform school into a life-changing experience for students.

The Innovator's Mindset

Empower Learning, Unleash Talent, and Lead a Culture of Creativity

By George Couros (@gcouros)

The traditional system of education requires students to hold their questions and compliantly stick to the scheduled curriculum. But our job as educators is to provide new and better opportunities for our students. It's time to recognize that compliance doesn't foster innovation, encourage critical thinking, or inspire creativity—and those are the skills our students need to succeed. In *The Innovator's Mindset*, George Couros encourages teachers and administrators to empower their learners to wonder, to explore—and to become forward-thinking leaders.

P is for PIRATE

Inspirational ABC's for Educators

By Dave and Shelley Burgess (@Burgess_Shelley)

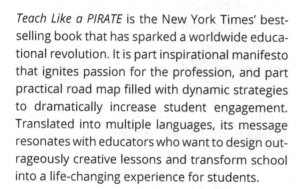

Teaching is an adventure that stretches the imagination and calls for creativity every day! In *P is for PIRATE*, husband and wife team Dave and Shelley Burgess encourage and inspire educators to make their classrooms fun and exciting places to learn. Tapping into years of personal experience and drawing on the insights of more than seventy educators, the authors offer a wealth of ideas for making learning and teaching more fulfilling than ever before.

Pure Genius

Building a Culture of Innovation and Taking 20% Time to the Next Level

By Don Wettrick (@DonWettrick)

For far too long, schools have been bastions of boredom, killers of creativity, and way too comfortable with compliance and conformity. In *Pure Genius*, Don Wettrick explains how collaboration—with experts, students, and other educators—can help you create interesting, and even life-changing, opportunities for learning. Wettrick's book inspires and equips educators with a systematic blueprint for teaching innovation in any school.

Learn Like a PIRATE

Empower Your Students to Collaborate, Lead, and Succeed

By Paul Solarz (@PaulSolarz)

Today's job market demands that students be prepared to take responsibility for their lives and careers. We do them a disservice if we teach them how to earn passing grades without equipping them to take charge of their education. In *Learn Like a PIRATE*, Paul Solarz explains how to design classroom experiences that encourage students to take risks and explore their passions in a stimulating, motivating, and supportive environment where improvement, rather than grades, is the focus. Discover how student-led classrooms help students thrive and develop into self-directed, confident citizens who are capable of making smart, responsible decisions, all on their own.

50 Things You Can Do with Google Classroom

By Alice Keeler and Libbi Miller (@alicekeeler, @MillerLibbi)

It can be challenging to add new technology to the classroom, but it's a must if students are going to be well-equipped for the future. Alice Keeler and Libbi Miller shorten the learning curve by providing a thorough overview of the Google Classroom App. Part of Google Apps for Education (GAfE), Google Classroom was specifically designed to help teachers save time by streamlining the process of going digital. Complete with screenshots, *50 Things You Can Do with Google Classroom* provides ideas and step-by-step instructions to help teachers implement this powerful tool.

Ditch That Textbook

Free Your Teaching and Revolutionize Your Classroom

By Matt Miller (@jmattmiller)

Textbooks are symbols of centuries-old education. They're often outdated as soon as they hit students' desks. Acting "by the textbook" implies compliance and a lack of creativity. It's time to ditch those textbooks—and those textbook assumptions about learning! In *Ditch That Textbook*, teacher and blogger Matt Miller encourages educators to throw out meaningless, pedestrian teaching and learning practices. He empowers them to evolve and improve on old, standard teaching methods. *Ditch That Textbook* is a support system, toolbox, and manifesto to help educators free their teaching and revolutionize their classrooms.

50 Things to Go Further with Google Classroom

A Student-Centered Approach

By Alice Keeler and Libbi Miller (@alicekeeler, @MillerLibbi)

Today's technology empowers educators to move away from the traditional classroom where teachers lead and students work independently—each doing the same thing. In *50 Things to Go Further with Google Classroom: A Student-Centered Approach*, authors and educators Alice Keeler and Libbi Miller offer inspiration and resources to help you create a digitally rich, engaging, student-centered environment. They show you how to tap into the power of individualized learning that is possible with Google Classroom.

140 Twitter Tips for Educators

Get Connected, Grow Your Professional Learning Network, and Reinvigorate Your Career

By Brad Currie, Billy Krakower, and Scott Rocco
(@bradmcurrie, @wkrakower, @ScottRRocco)

Whatever questions you have about education or about how you can be even better at your job, you'll find ideas, resources, and a vibrant network of professionals ready to help you on Twitter. In *140 Twitter Tips for Educators*, #Satchat hosts and founders of Evolving Educators Brad Currie, Billy Krakower, and Scott Rocco offer step-by-step instructions to help you master the basics of Twitter, build an online following, and become a Twitter rock star.

The Zen Teacher

Creating Focus, Simplicity, and Tranquility in the Classroom

By Dan Tricarico
(@thezenteacher)

Teachers have incredible power to influence—even improve—the future. In *The Zen Teacher*, educator, blogger, and speaker Dan Tricarico provides practical, easy-to-use techniques to help teachers be their best—unrushed and fully focused—so they can maximize their performance and improve their quality of life. In this introductory guide, Dan Tricarico explains what it means to develop a Zen practice—something that has nothing to do with religion and everything to do with your ability to thrive in the classroom.

Master the Media

How Teaching Media Literacy Can Save Our Plugged-in World

By Julie Smith (@julnilsmith)

Written to help teachers and parents educate the next generation, *Master the Media* explains the history, purpose, and messages behind the media. The point isn't to get kids to unplug; it's to help them make informed choices, understand the difference between truth and lies, and discern perception from reality. Critical thinking leads to smarter decisions—and it's why media literacy can save the world.

eXPlore Like a Pirate

Gamification and Game-Inspired Course Design to Engage, Enrich, and Elevate Your Learners

By Michael Matera
(@MrMatera)

Are you ready to transform your classroom into an experiential world that flourishes on collaboration and creativity? Then set sail with classroom game designer and educator Michael Matera as he reveals the possibilities and power of game-based learning. In *eXPlore Like a Pirate*, Matera serves as your experienced guide to help you apply the most motivational techniques of gameplay to your classroom. You'll learn gamification strategies that will work with and enhance (rather than replace) your current curriculum and discover how these engaging methods can be applied to any grade level or subject.

The Classroom Chef

Sharpen Your Lessons. Season Your Classes. Make Math Meaningful.

By John Stevens and Matt Vaudrey
(@Jstevens009, @MrVaudrey)

In *The Classroom Chef*, math teachers and instructional coaches John Stevens and Matt Vaudrey share their secret recipes, ingredients, and tips for serving up lessons that engage students and help them "get" math. You can use these ideas and methods as-is, or better yet, tweak them and create your own enticing educational meals. The message the authors share is that, with imagination and preparation, every teacher can be a Classroom Chef.

The Writing on the Classroom Wall

How Posting Your Most Passionate Beliefs about Education Can Empower Your Students, Propel Your Growth, and Lead to a Lifetime of Learning

By Steve Wyborney
(@SteveWyborney)

In *The Writing on the Classroom Wall*, Steve Wyborney explains how posting and discussing Big Ideas can lead to deeper learning. You'll learn why sharing your ideas will sharpen and refine them. You'll also be encouraged to know that the Big Ideas you share don't have to be profound to make a profound impact on learning. In fact, Steve explains, it's okay if some of your ideas fall *off* the wall. What matters most is sharing them.

How Much Water Do We Have?

5 Success Principles for Conquering Any Change and Thriving in Times of Change

By Pete Nunweiler with Kris Nunweiler

In *How Much Water Do We Have?* Pete Nunweiler identifies five key elements—information, planning, motivation, support, and leadership—that are necessary for the success of any goal, life transition, or challenge. Referring to these elements as the 5 Waters of Success, Pete explains that, like the water we drink, you need them to thrive in today's rapidly paced world. If you're feeling stressed out, overwhelmed, or uncertain at work or at home, pause and look for the signs of dehydration. Learn how to find, acquire, and use the 5 Waters of Success—so you can share them with your team and family members.

Kids Deserve It!

Pushing Boundaries and Challenging Conventional Thinking

By Todd Nesloney and Adam Welcome
(@TechNinjaTodd, @awelcome)

In *Kids Deserve It!*, Todd and Adam encourage you to think big and make learning fun and meaningful for students. Their high-tech, high-touch, and highly engaging practices will inspire you to take risks, shake up the status quo, and be a champion for your students. While you're at it, you just might rediscover why you became an educator in the first place.

LAUNCH

Using Design Thinking to Boost Creativity and Bring Out the Maker in Every Student

By John Spencer and A.J. Juliani (@spencerideas, @ajjuliani)

Something happens in students when they define themselves as *makers* and *inventors* and *creators*. They discover powerful skills—problem-solving, critical thinking, and imagination—that will help them shape the world's future … *our* future. In *LAUNCH*, John Spencer and A.J. Juliani provide a process that can be incorporated into every class at every grade level … even if you don't consider yourself a "creative teacher." And if you dare to innovate and view creativity as an essential skill, you will empower your students to change the world—starting right now.

Your School Rocks … So Tell People!

Passionately Pitch and Promote the Positives Happening on Your Campus

By Ryan McLane and Eric Lowe (@McLane_Ryan, @EricLowe21)

Great things are happening in your school every day. The problem is, no one beyond your school walls knows about them. School principals Ryan McLane and Eric Lowe want to help you get the word out! In *Your School Rocks … So Tell People!* McLane and Lowe offer more than seventy immediately actionable tips along with easy-to-follow instructions and links to video tutorials. This practical guide will equip you to create an effective and manageable communication strategy using social media tools. Learn how to keep your students' families and community connected, informed, and excited about what's going on in your school.

Play Like a Pirate

Engage Students with Toys, Games, and Comics

By Quinn Rollins (@jedikermit)

Yes! School can be simultaneously fun and educational. In *Play Like a Pirate*, Quinn Rollins offers practical, engaging strategies and resources that make it easy to integrate fun into your curriculum. Regardless of the grade level you teach, you'll find inspiration and ideas that will help you engage your students in unforgettable ways.

Instant Relevance

Using Today's Experiences in Tomorrow's Lessons

By Denis Sheeran (@MathDenisNJ)

Every day, students in schools around the world ask the question, "When am I ever going to use this in real life?" In *Instant Relevance*, author and keynote speaker Denis Sheeran equips you to create engaging lessons *from* experiences and events that matter to your students. Learn how to help your students see meaningful connections between the real world and what they learn in the classroom—because that's when learning sticks.

Escaping the School Leader's Dunk Tank

How to Prevail When Others Want to See You Drown

By Rebecca Coda and Rick Jetter (@RebeccaCoda, @RickJetter)

No school leader is immune to the effects of discrimination, bad politics, revenge, or ego-driven coworkers. These kinds of dunk tank situations can make an educator's life miserable. By sharing real-life stories and insightful research, the authors (who are dunk tank survivors themselves) equip school leaders with the practical knowledge and emotional tools necessary to survive and, better yet, avoid getting "dunked."

Lead Like a PIRATE

Make School Amazing for Your Students and Staff

By Shelley Burgess and Beth Houf (@Burgess_Shelley, @BethHouf)

In *Lead Like a PIRATE*, education leaders Shelley Burgess and Beth Houf map out the character traits necessary to captain a school or district. You'll learn where to find the treasure that's already in your classrooms and schools—and how to bring out the very best in your educators. This book will equip and encourage you to be relentless in your quest to make school amazing for your students, staff, parents, and communities.

Start. Right. Now.

Teach and Lead for Excellence

By Todd Whitaker, Jeff Zoul, and Jimmy Casas (@ToddWhitaker, @Jeff_Zoul, @casas_jimmy)

In their work leading up to *Start. Right. Now.* Todd Whitaker, Jeff Zoul, and Jimmy Casas studied educators from across the nation and discovered four key behaviors of excellence: Excellent leaders and teachers *Know the Way, Show the Way, Go the Way, and Grow Each Day*. If you are ready to take the first step toward excellence, this motivating book will put you on the right path.

Table Talk Math

A Practical Guide for Bringing Math into Everyday Conversations

By John Stevens (@Jstevens009)

Making math part of families' everyday conversations is a powerful way to help children and teens learn to love math. In *Table Talk Math*, John Stevens offers parents (and teachers!) ideas for initiating authentic, math-based conversations that will get kids notice and be curious about all the numbers, patterns, and equations in the world around them.

About the Authors

Photo by Alex Kang / heyitsalex.com

Alice Keeler is a mom of five children. She taught high school math for fourteen years and currently teaches in the credential program at California State University Fresno. She is a Google Certified Innovator and frequently blogs on teaching with Google Apps at alicekeeler.com. Alice co-authored the books *50 Things You Can Do With Google Classroom, 50 Things to Go Further with Google Classroom: A Student-Centered Approach, Google Apps for Littles,* and *Ditch that Homework.*

Diana Herrington is a member of National Council of Teachers of Mathematics (NCTM), National Science Teachers Association (NSTA), California Mathematics Council (CMC), Computer Using Educators (CUE), and has served on numerous state committees on assessment and teaching task forces, along with being a member of the California Teacher Advisory Council (CalTAC).

Diana has been recognized for her teaching with the following awards: Presidential Awardee for Excellence in Mathematics and Science Teaching (PAEMST), the first non-science teacher to receive the California State Science Fair Coach of the Year, and Central Valley Computer Using Educator of the Year (CVCUE).

Upon retiring from twenty-nine years of teaching mathematics at Clovis High School, she began teaching part time at California State University, Fresno. At CSUF she works with pre-service teachers at the beginning of their teaching journey.

CPSIA information can be obtained
at www.ICGtesting.com
Printed in the USA
LVHW061537230720
661378LV00002B/30